SpringerBriefs in Applied Sciences and Technology

PoliMI SpringerBriefs

Springer, in cooperation with Politecnico di Milano, publishes the PoliMI Springer-Briefs, concise summaries of cutting-edge research and practical applications across a wide spectrum of fields. Featuring compact volumes of 50 to 125 (150 as a maximum) pages, the series covers a range of contents from professional to academic in the following research areas carried out at Politecnico:

- Aerospace Engineering
- Bioengineering
- Electrical Engineering
- Energy and Nuclear Science and Technology
- Environmental and Infrastructure Engineering
- Industrial Chemistry and Chemical Engineering
- Information Technology
- Management, Economics and Industrial Engineering
- Materials Engineering
- Mathematical Models and Methods in Engineering
- Mechanical Engineering
- Structural Seismic and Geotechnical Engineering
- Built Environment and Construction Engineering
- Physics
- Design and Technologies
- Urban Planning, Design, and Policy

http://www.polimi.it

Giovanni Lanza

Enabling Immobilities: Social and Spatial Implications for Urban Planning

POLITECNICO
MILANO 1863

Giovanni Lanza
DAStU
Politecnico di Milano
Milan, Italy

ISSN 2191-530X   ISSN 2191-5318 (electronic)
SpringerBriefs in Applied Sciences and Technology
ISSN 2282-2577   ISSN 2282-2585 (electronic)
PoliMI SpringerBriefs
ISBN 978-3-031-80998-9   ISBN 978-3-031-80999-6 (eBook)
https://doi.org/10.1007/978-3-031-80999-6

This Springer imprint is published by the registered company Springer Nature Switzerland AG
The registered company address is: Gewerbestrasse 11, 6330 Cham, Switzerland

If disposing of this product, please recycle the paper.

Acknowledgments

The author wishes to acknowledge that part of the research conducted for the writing of this book was developed within and funded by the Horizon 2020 project *PoliVisu: Policy Development Based on Advanced Geospatial Data Analytics and Visualisation* (Grant Agreement ID: 769608) and the project *EX-TRA: Experimenting with City Streets to Transform Urban Mobility*, funded by the JPI Urban Europe ERA-NET Cofund Urban Accessibility and Connectivity (ENUAC).

The author would like to express heartfelt thanks to Professors Paola Pucci and Giovanni Vecchio for their guidance, as well as to Professors Luca Bertolini, Matteo Colleoni, Grazia Concilio, Luca Staricco for their valuable contributions to this research.

Gratitude is also extended to colleagues Fabio Manfredini and Carmelo Di Rosa from the Maudlab laboratory at the Politecnico di Milano for their assistance in collecting and analyzing the mobile phone data used in Chap. 3. Special thanks go to Dr. Bruna Vendemmia for her collaboration during the fieldwork conducted in the Piacenza Apennines area, whose findings are discussed in Chap. 5.

Contents

Chapter 1
Introduction

Abstract The introduction establishes the theoretical positioning of the book, concisely outlining the perspective through which the concept of immobility will be examined in greater depth throughout the subsequent chapters. Following a detailed explanation of the rationale for a theoretical and practical focus on immobility within the fields of urban studies and planning, the chapter proceeds to present the overall structure of the book. In doing so, it directly engages with four key issues central to the analysis of immobility, which must be addressed to explore its social and spatial implications. These issues will be further scrutinized through theoretical and empirical investigation in the following chapters. In its conclusion, the introduction provides an overview of the potential impacts of a more comprehensive understanding of immobility and its socio-spatial implications, both at the scientific and practical levels, as well as within broader societal contexts.

1.1 The Relevance of a Focus on Immobility

Addressing immobility in the territories of contemporary cities represents a significant theoretical and operational challenge, made pertinent by various contingencies and the complex crises that current society faces, starting with the climate emergency. The events stemming from the outbreak of the Covid-19 pandemic have placed immobility—understood as a differently lasting slowdown or interruption of our daily mobility practices—at the center of cultural and political debate, as well as in people's daily lives. Indeed, the Pandemic had represented, for some months at least, a moment when most people had direct experience of some form of immobility. Certainly, physical immobility, with people's bodies confined within increasingly restricted spaces following the rules implemented to contain the virus' spreading. But immobility has also took on different forms, linked to the perception of unprecedented (at least for some part of the global population) limits in displacing in a world that perceives movement as an essential factor in positioning oneself in society and in establishing social relations. The main consequence for many people was an increasingly strong

G. Lanza, *Enabling Immobilities: Social and Spatial Implications for Urban Planning*,
PoliMI SpringerBriefs, https://doi.org/10.1007/978-3-031-80999-6_1

and distressing impression of the growing number of missed opportunities and unexplored possibilities induced by the inability to move. Yet, the direct experience of this condition imposed on many has led to a renewed interest towards the rediscovery of the meaning and value of staying still or moving less than before or others. For some individuals, this process has prompted a reconsideration of their own rhythms and mobility practices, leading to behavioral changes or the development of new sensitivities that persist even today, after the pandemic. The pandemic has also had significant effects in the field of urban planning, reinforcing an increased awareness of the importance of proximity, accessibility, and the need to reduce the environmental and social impacts produced by the desire and need to move. This awareness is not entirely new, as illustrated in the book, and is widely inspired by previously developed and well-established planning models and concepts. However, the challenges humanity faces today, starting with the climate emergency and the relentless process of global urbanization, make it urgent to reflect more deeply on immobility and the contribution that planning for immobility could make in designing more inclusive and sustainable cities and territories of tomorrow where the need for extensive, resource-consuming, and costly travel could—and should—be reduced. Such a contribution must be explored both theoretically, to identify a conceptual framework useful for analyzing immobility and its socio-spatial implications, and operationally, towards the construction of methodologies and analytical tools that can support the implementation of fair and sustainable immobility policies.

These challenges are addressed within this book which, drawing inspiration from the extensive theoretical literature analyzing human spatial mobility, conversely emphasizes the importance of focusing on immobility, a focus that may produce substantial implications for people, urban studies, and planning. One way to investigate the relevance of immobility—a condition that, unlike mobility, is often under examined and generally neglected in the academic debate (Schewel 2020)—is to start from the role and importance that literature from various disciplines and, more broadly, societies and individuals attribute to mobility. This central role has been explained by considering that mobility does not only end in simple movement in space—a dimension that is still central to our interpretation, especially in urban studies—but that is increasingly conceptualized as a set of meaningful social practices that make up social, cultural, and political life (Adey et al. 2014, p.3). Being mobility a pervasive and fundamental aspect of our human experience, it becomes central to both the world and our understanding of it, able to assuming multiple forms and meanings and producing (and, in turn, being shaped by) social relations (Cresswell 2006, 2010). Indeed, mobility is both seen as a spacetime expression of the unequal power relations that structure society according to ethnic, gender, demographic, economic and educational status lines (Sheller and Urry 2006; Sheller 2014) and a key resource for social inclusion, representing a basic social and spatial capital which is essential to facilitating interaction and activity participation (Lucas 2012; Martens 2017). However, because these unequal power dynamics influence its distribution, mobility is a resource that is not equally accessible to all (Cresswell 2010).

In a world that increasingly values mobility as a fundamental enabler of social interaction and associates it with positive outcomes in constructing social status and personal fulfilment (Elliott and Urry 2010), the ability and willingness to move significantly shape an individual's social role potentially disadvantaging those with less access to this resource (Kaufmann et al. 2004). Hence, the study of mobility practices has clear political implications: it becomes an essential prerequisite for understanding the spatial and social factors that facilitate and support the movement of some and, on the other hand (and simultaneously), inhibit, limit or prevent it for others, contributing to the emergence of what we can refer to as *mobility differentials* between individuals. Mobility differentials can be interpreted as the different ways people appropriate and act their mobilities according to their possibilities, abilities and personal desires shaped by the socio-spatial context of belonging. The research featured in this book interprets mobility differentials as the varying intensities of individuals' daily mobility practices in space and time. It follows that immobility is not referred to as absolute but is considered the result of a choice or constraint that makes some people less likely to move, thus making them less mobile—or relatively immobile—than others (Adey 2006).

Considering the importance of mobility in everyday life and for social inclusion and participation, it could be hypothesized that mobility differentials producing relative immobilities may uniquely stem from specific factors that, acting as barriers, prevent or inhibit a person's movements. This interpretation implies a binary relationship between mobility/immobility and social privilege/exclusion, an assumption that would be capable of explaining the emergence of a society of enclaves (Turner 2007), where mobility becomes the interpretative key to shed light on unequal power relations and their evolution over time. In this view, immobility would be considered a 'negative' phenomenon, the product of an imposition determined by an unequal distribution of the resource of mobility linked to social, territorial and infrastructural imbalances which can cause and result from social exclusion and lack of participation.

However, to show that immobility, like mobility, can be a complex and multifaceted phenomenon, a limited propensity for mobility can also be conceived as the result of individual behaviors influenced by a voluntary narrowing of the spatial extent of one's daily activity area and the time spent in movement. Some individuals may move less than others by choice rather than constraint, and this may also occur due to peculiar socio-spatial configurations of their living environments. For instance, when necessary and desired services, activities, and social ties are within close physical proximity, individuals can rationalize and save different resources related to mobility. Among these, all the natural resources that are exploited to keep transport systems active and support our increasingly complex and spatially fragmented individual mobility practices, or the economic and temporal resources required to move. Also, new forms of "chosen" immobility may be induced by emerging models of organization of work and daily life connected to the development of digital communication technologies.

Thus, relative immobility, if paired with personal and contextual features that allow access to goods and resources in virtual and physical proximity, may not compromise an individual's level and perception of inclusion and participation. This situation can be termed *reversible* relative immobility, which is opposed to all those *constrained* conditions where the absence of mobility is not the result of a voluntary choice but is determined by barrier factors limiting the possibilities for movement and participation. Conversely—and this is an element that is progressively explored in the book—mobility itself can also be read according to the same approach.

A crucial role in this conceptualization is played by the concepts of proximity and accessibility, where the latter is here conceived as a measure of the individual's capability to reach and engage in activities—also in virtual form—considered valuable for their daily lives, thus emphasizing the dimensions of potential and personal abilities rather than only focusing on the provision, availability and efficiency of transport systems (Kenyon et al. 2002; Geurs and van Wee 2004; Jones and Lucas 2012, p. 6). In fact, good levels of accessibility by proximity, or the possibility to reach physically proximate activities and opportunities without extended travel in time and space, can be crucial in inducing reversible immobility by ensuring individuals' activity participation (Lucas 2012; Martens 2017) even with low mobility and travel. Conversely, an individual's relative immobility can have adverse social outcomes if not counterbalanced by virtual availability, physical proximity or the ability to reach the resources a person requires. While not being causally linked to the emergence of specific forms of mobility and immobility, accessibility by proximity is a promising concept of reference to understand the possible social and spatial implications of various forms of relative immobilities, especially in planning, where a theoretical and operative debate on accessibility and its measurement is well established.

The proposed conceptualization of immobility, which represents the basis of the book's theoretical structure and framework, can also affect how conditions, experiences and implications of immobility are analyzed and addressed through planning policies. For example, in the case of constrained immobility, this perspective can help design integrated land use and transport measures to enhance accessibility and social inclusion for disadvantaged individuals, populations and territories. For reversible immobility, this framework can help detect the conditions that guarantee accessibility by proximity bringing people to live conditions of immobility in a positive way to be replicated in other contexts. This last point introduces the notion of immobility as a form of social capital based on the existence and maintenance of social and spatial relations of proximity in hypothetical scenarios in which high levels of high-frequency mobility over long distances would be no longer essential. In this view, immobility is proposed by some authors (see, for instance, Ferreira et al. 2017) as a form of resilience against the economic, social and environmental unsustainability of the increasingly complex and vulnerable contemporary hyper-mobility systems: according to these authors, a greater focus on immobility (why, where, when and for whom it occurs), is needed to ensure an inclusive and more sustainable future mobility (Lucas and Madre 2018) and help solve the resilience issue of hyper-mobility produced by its increasing costs and failures, which are increasingly evident today even after the temporary forced pause caused by the Covid-19 pandemic.

Following this crucial suggestion, the book aims to provide a conceptual and operational analytical approach to immobility within transport and land use planning, offering new insights for inclusive and sustainable planning policies.

1.2 Structure of the Book

The book proposes an exploration of immobility—understood as a concept worth deepening and as an experienced human condition—and provides theoretical and practical insights on how to analyze it.

The starting point, presented in Chap. 2, is a review of urban studies and planning literature that addresses immobility, aiming to construct a framework for a theoretical and operative exploration of its spatial and social implications. In its conclusions, the chapter delineates four relevant issues this exploration must address.

The first issue concerns the measurement of immobility, understood as an essential activity to quantify the extent of mobility differentials existing between different individuals and social groups in specific spatial settings. This topic is addressed in Chap. 3, which, drawing from the plurality of available quantitative and qualitative tools and methods to capture and reconstruct complex sets of realized mobility and immobility practices, identifies opportunities and limitations of these approaches and explores new research avenues enabled by the availability and use of digital data for urban and territorial research. Thus, an empirical exploration is proposed to measure immobility patterns in a specific territorial context (the Piacenza Apennine inner area, a low density and shrinking mountain area in northern Italy) using mobile phone data. The results of this experimentation, the utility of the approach, and possible future research avenues are discussed in the chapter's conclusion.

The second issue concerns the assessment of the spatial conditions enabling chosen immobility, comprehensively addressed in the book's fourth chapter. Directly referring to the theoretical framework illustrated in Chap. 2, this section focuses on accessibility by proximity measurement as an essential analytical approach to understanding under what spatial conditions immobility can be configured as a choice or a constraint. Indeed, immobility can be enabled by ensuring access through short-range, sustainable mobility options to a set of daily relevant services in diverse territorial contexts. Such conditions can only be promoted through local planning and transport policies designed with the support of highly detailed and context-sensitive accessibility assessment tools. The chapter provides an overview of the main challenges related to accessibility by proximity measurements for immobility enablement, identifying some critical points specifically concerning the definition of activities to be included in the measurement, the transport modes considered, and the user profiles included in the simulation The empirical section of the chapter describes the methods and results of an experimental accessibility index applied in the territorial context of the Piacenza Apennines that copes with several measurement challenges, opening up new avenues for accessibility analysis and policy-related considerations about the promotion of immobility especially in peri-urban territories.

The third issue, addressed in the fifth chapter, concerns the analysis of the experience of immobility, understood as the exploration of the complex variety of lived experiences of immobility at the individual and collective level and the reasons behind this choice or constraint. The chapter employs the technique of microstories to investigate, in the case study of the Piacenza Apennine, the individual causes and effects of immobility in relation to local accessibility conditions. This approach enriches the understanding of immobility featured in the theoretical framework proposed in the second chapter, showing how immobility and mobility can simultaneously be conditions that may be conceived as chosen or imposed depending on the interrelationship between contextual factors and personal experiences. Furthermore, the possibility is discussed that specific individual or contextual factors, especially those induced by the outcomes of a combination of public policies and community collaboration, may contribute to reversing constrained conditions toward chosen mobility and immobility. The concepts of mobility and immobility enablement are thus introduced to outline the possible roles of public policies and community action in promoting and favoring accessibility by proximity where possible (immobility enablement) or inclusive and sustainable mobility options (mobility enablement) where accessibility by proximity cannot be guaranteed.

The fourth and final issue, addressed in the sixth chapter, concerns the promotion of chosen, reversible immobility and mobility through integrated planning and transport policies. Making reference to the conclusions of the previous chapter, the text is organized into two main sections. The first refers to policies for immobility enablement aimed at improving the direct or indirect provision of activities and services that can generate forms of functional and relational proximity. The second part refers to mobility enablement, focusing on possible initiatives to support sustainable intermobility favored by the development of new technologies in telecommunication and transport. Both parts reference examples of best practices policies that can be applied in various settlement contexts to extend models and visions traditionally associated with compact cities to low-density contexts as well.

1.3   The Potential Impacts of a Deeper Understanding of Immobility

From a scientific perspective, this book contributes to deepening the understanding of immobility, which refers to a multifaceted condition with different social and spatial causes and outcomes often neglected in scientific literature. The interest in these forms and conditions has emerged, at the theoretical level, mainly following the advent of the so-called new mobilities paradigm and the progressive recognition of mobility's social, cultural, and political relevance. A discourse on immobilities has developed as a natural consequence of this interest, leading some authors to question the nexus that links forms and moments of mobility and immobility (Adey 2006) and the social and political reasons underlying the mobility differentials that characterize distinct individuals. This perspective has originated interesting conceptualizations that are referenced, discussed, and partly questioned in this book. On the contrary, urban and transport planning practices have rarely addressed a perspective on immobility that considers and distinguishes situations in which this condition can be traced back to a social issue from others that embody an opportunity for designing policies and orienting urban and social development to promote social inclusion and environmental sustainability. This work emancipates the concept of immobility, discussing different perspectives and defining a theoretical position that bridges theory and practice across disciplines, targeting a diverse audience. It is intended for academic researchers as well as for various stakeholders, including planning practitioners, professionals, and public authorities.

The same theoretical position shapes a framework for analysing immobility between theory and practice that becomes the main element explaining the practical relevance of the research: throughout its chapters, the book identifies a set of relevant issues concerning the understanding and measurement of immobility and its social and spatial implications proposing a set of analytical quantitative and qualitative techniques that are discussed, reinterpreted and empirically tested to emphasize their relevance in exploring forms of immobility-related disadvantages and opportunities, and in increasing the awareness on the policy value of such exploration. Not only does a focus on immobility allow relating the mobility behaviors of individuals to the specific social and spatial characteristics of their living context, but also drawing essential insights on the possibilities of increasing the opportunities for activity participation and accessibility, not solely relying on the enhancement of travel and mobility. Also, the decision to apply these techniques to the same territorial case study (the Piacenza Apennine area) also allows for identifying the possibilities of integrating and combining these different techniques, revealing the analytical and policy-related value of a mixed-methods approach.

In this perspective the social relevance of the research is expressed in the proposal of possible interventions and policy measures oriented by the analytical methodologies towards the promotion of forms of local development. In these actions, inspired by the concept of accessibility by proximity—that is increasingly central to the planning debate also as a result of the effects of the Covid-19 pandemic—promoting forms of relative immobility may become an actual policy goal to imagine more equitable and environmentally sustainable cities and territories, especially, as explored in the book, if inspired by how people already construct their immobilities by responding and reversing constrained conditions.

References

Adey P (2006) If mobility is everything then it is nothing: Towards a relational politics of (im)mobilities. Mobilities 1(1):75–94. https://doi.org/10.1080/17450100500489080

Adey P, Bissell D, Hannam K, Merriman P, Sheller M (2014) Introduction at the Routledge handbook of mobilities. In: Adey P, Bissell D, Hannam K, Merriman P, Sheller M (Eds.) The Routledge handbook of mobilities, Routledge, Abingdon

Cresswell T (2006) On the move: mobility in the modern western world. Routledge, London

Cresswell T (2010) Towards a politics of mobility. Environ Plan d: Soc Space 28(1):17–31. https://doi.org/10.1068/d11407

Elliott A, Urry J (2010) Mobile lives. Routledge, New York

Ferreira A, Bertolini L, Næss P (2017) Immotility as resilience? A key consideration for transport policy and research. Applied Mobilities 2(1):16–31. https://doi.org/10.1080/23800127.2017. 1283121

Geurs KT, van Wee B (2004) Accessibility evaluation of land-use and transport strategies: review and research directions. J Transp Geogr 12(2):127–140. https://doi.org/10.1016/j.jtrangeo.2003. 10.005

Jones P, Lucas K (2012) The social consequences of transport decision-making: clarifying concepts, synthesising knowledge and assessing implications. J Transp Geogr 21:4–16. https://doi.org/10. 1016/j.jtrangeo.2012.01.012

Kaufmann V, Bergman MM, Joye D (2004) Motility: mobility as capital. Int J Urban Reg Res 28(4):745–756

Kenyon S, Lyons G, Rafferty J (2002) Transport and social exclusion: Investigating the possibility of promoting inclusion through virtual mobility. J Transp Geogr 10(3):207–219

Lucas K (2012) Transport and social exclusion: where are we now? Transp Policy 20:105–113

Lucas K, Madre JL (2018) Workshop synthesis: dealing with immobility and survey non-response. Transp Res Procedia 32:260–267. https://doi.org/10.1016/j.trpro.2018.10.048

Martens K (2017) Transport justice. Designing fair transportation systems. Routledge, Abingdon

Schewel K (2020) Understanding immobility: moving beyond the mobility bias in migration studies. Int Migr Rev 54(2):328–355. https://doi.org/10.1177/0197918319831952

Sheller M (2014) The new mobilities paradigm for a live sociology. Curr Sociol 62(6):789–811. https://doi.org/10.1177/0011392114533211

Sheller M, Urry J (2006) The new mobilities paradigm. Environ Plan A 38(2):207–226. https://doi. org/10.1068/a37268

Turner, BS (2007) The enclave society: Towards a sociology of immobility. Eur J Soc Theory 10(2):287–304. https://doi.org/10.1177/1368431007077807

Chapter 2
Immobility: Choice or Constraint?

Abstract The chapter develops a theoretical framework underscoring the importance of immobility as a critical subject for study and analysis through an extensive review of literature spanning multiple disciplines. This positioning synthesizes a range of viewpoints and perspectives, shaped by ongoing debates that consider the relative absence of mobility (immobility) as a condition with profound social and spatial implications. Furthermore, the chapter examines how this phenomenon and its associated practices can relate to different levels of accessibility and proximity to valuable spatial opportunities, highlighting the variability of this relationship across different spatial and social contexts. This approach introduces a novel framework that bridges theory and practice, conceptualizing immobility as a result of choice, constraint, or a combination of both. The chapter concludes by identifying four critical methodological and policy-related issues essential for the analysis of immobility, which are crucial for its application in urban studies and planning.

2.1 Introduction

The concept of spatial immobility and the in-depth analysis of the social, cultural, and geographical characteristics influencing this condition are often examined, in scientific literature, focusing primarily on mobility and travel. This approach is unsurprising since, by definition, immobility indicates an absence of movement, a state of stasis interrupting or opposing the flows of generalized mobility that characterize contemporary societies (Bauman 1998; Urry 2000, 2005; Bourdin 2005; Le Breton 2005; Elliott and Urry 2010). These flows, consisting of complex combinations of physical movements, representations and practices across time and space (Cresswell 2010), are the subject of a constantly expanding field of research. Such centrality of the topic in various academic fields is primarily due to the role of mobility in ensuring participation in social life, access to resources, opportunities, and activities distributed in space, and the pursuit of individual and collective aspirations based on personal capabilities and the possibilities offered by the context of belonging (Sen 1992; Geurs and van Wee 2004; Kaufmann et al. 2002, 2004; Farrington et al.

2005; Sheller and Urry 2006; Vannini 2010; Martens et al. 2014; Pucci and Vecchio 2019). Indeed, rather than being only considered a matter of transport offer, mobility is increasingly intended as an essential component of social life (Cresswell 2006; Lucas 2012; Lucas et al. 2016; Martens 2017).

Since mobility assumes such a significant role in our human experience, it is easy to understand, in contrast, the relevance of a focus on the conditions of immobility, namely all those situations in which the spatial movements of people and things are absent, arrested or limited for different reasons and with different meanings and effects. If Cresswell (2010) explains movement as the result of an internal or external force applied to an individual or object, immobility can similarly be considered a condition where a force, either voluntarily exerted or externally imposed, interrupts or inhibits spatial movement. Academic interest has thus developed around the social, spatial, cultural, and political reasons behind this force and its consequences on the same spheres (Hannam et al. 2006; Perkins 2019).

The reasons and effects of immobility are thus intertwined with those of mobility, creating a varied and complex picture constituted by the multiple ways people move—or do not move (Adey 2006). This complexity suggests considering immobilities in plural terms and opens the field to diverse interpretations found in the literature about the dynamic, multifaceted, and often ambiguous relationship between mobility and immobility. These interpretations are naturally influenced by cultural and disciplinary references guiding the analysis and exploration of these themes. Therefore, introducing a holistic definition of immobility accounting for all its facets is neither easy nor necessarily helpful. Instead, this chapter aims to provide a comprehensive overview of the various perspectives through which immobility is conceived, studied, and explored in the literature. This focus is essential to establish a theoretical and operational framework proposed in this book.

The literature review reveals that interest in immobility and the analysis of its characteristics, underlying reasons, and meanings at both individual and collective levels has gained momentum with the rise of the new mobilities paradigm. This paradigm shift has led to new perceptions and interests in the conditions underpinning the generation and experience of movement and stillness and their broader impacts across various socio-cultural, economic, and political contexts (Shaw and Hesse 2010). Consequently, the in-depth study of immobility is a natural extension of mobilities studies, with much of the existing research employing methods and objectives from disciplines such as sociology, geography, anthropology, and migration studies.

Conversely, the notion of immobility is still used only marginally in urban studies and planning. This lack of interest may be due to the traditional approach of transport studies, which rely on quantifying mobility demand to design, plan, and justify the implementation of efficient transport systems (Rodrigue et al. 2016). The consequence is the usual overlooking of social disadvantage and exclusion deriving from missed mobility opportunities, while the focus on needs in transport modeling tends to structurally favor transport improvements for already mobile groups (Martens 2006; van Wee and Geurs 2011; Pucci and Vecchio 2019). Connected to this approach is the idea that forms of immobility are primarily associated with physical and social stasis leading to deprivation and exclusion, reflecting the inability or impossibility to actively participate and access social networks and spatial activities and opportunities (Stanley and Vella-Brodrick 2011). However, moving from a focus on actual mobility to one considering situations of unrealized movements together with a better understanding of the possible sociospatial reasons behind these events can be a promising way to update and broaden the scope of the analysis and tools proper for urban studies and planning moving beyond mobility-centered approaches.

An initial in-depth study of the complex and ambiguous relationship between mobility and immobility, interpreted through a relational and differential logic, forms the basis of this discussion (Sect. 2.2). This first approach lays the foundations for reflecting on the meaning of forms of relative immobility in a world that strongly depends on high-frequency, high-speed and high-distance mobilities. The result is the problematization of the link between immobility and conditions of social exclusion towards a more articulated reading (Sect. 2.3) that favors the interpretation of forms of low mobility allowed by accessibility by proximity as an opportunity to rethink the shapes, uses and organization of spaces and urban networks, especially in the aftermath of the Covid-19 pandemic. (Sect. 2.4). The theoretical content is then translated into a framework between theory and practice associating immobility and accessibility as the basis for operationalizing immobility within urban and transport planning, as presented in Sect. 2.5. The chapter concludes by outlining four critical issues related to immobility analysis that lay the foundation for the subsequent chapters of the book.

2.2 The Multifaceted Relationship Between Mobility and Immobility

From a conceptual standpoint, the literature focuses on immobility by exploring the possible links that bind this condition to various and contextual mobility practices. In this sense, reference is made to the definition proposed by Adey (2006) and inspired by Doel (1999). The authors conceive mobility and immobility as symbiotic phenomena that coexist, alternate and blend continuously, becoming elements of a relational continuum in which immobility manifests itself as a result of mobility differentials both at the individual and societal levels (Adey 2006; Pellegrino 2011; Motte-Baumvol et al. 2015). The relationship between mobility and immobility can thus be investigated from a dual interpretive key, namely in relational and differential terms.

The first interpretation (relational terms) focuses on the continuous alternations between mobility and immobility and their interdependence, as effectively proposed by Hannam et al. (2006). In their conceptualization, immobilities are considered as moments of anchorage—called moorings—acting as essential supports to the implementation and continuous regeneration of movement. According to the same authors, anything mobile needs to relate continuously to these systems of immobility to function correctly (Urry 2003), even in a society interpreted as liquid, network-based and deterritorialized in which everything is mobile (Bauman 2000; Urry 2005) and that glamorizes the growth of mobility and speed as key factors for the achievement and maintenance of privileged social statuses (Cohen and Gössling 2015). Further developing the concept, moorings can be attributed both a physical-material and an experiential value. In the first case, reference is made to the existence of spaces and devices that allow or force the anchoring of individual movements, such as infrastructures or transit spaces designed to offer, with their apparent static nature, an essential basis on which different trajectories and flows of people and objects can coexist and interface during moments of more or less temporary interruption (Bissell et al. 2011), generating a structural and natural intertwining between mobile and immobile spaces and times. Similarly, while certain types of physical moorings facilitate mobility for some, others may represent points of forced disruption of mobility flows, as is the case of physical or digital barriers in place to limit and prevent undesired migratory mobilities (Turner 2007; Mountz 2010). Conversely, the experiential value refers to the set of individual and social factors that are crucial in shaping the relationship and alternation between mobilities and immobilities in a process that can be observed in its short- and long-term effects. In the short term, we can notice how, in our daily experience, pauses and moments of stillness and anchorage underpin mobility experiences and are naturally required to rest, relate to the surrounding space and the people we encounter, or subsequently initiate other movements and practices. Extending this perspective to the long term, we can observe how an individual's life is punctuated by moments of greater or lesser propensity for mobility, to be read in relation to the capacity of each individual to plan and act on their mobilities following intentions, competences, wills, and needs, or to adapt to external rhythms and the mobilities of others. At the individual level, these capacities develop and change throughout life and are at least partly shaped by the direct experiences of socialization through which we internalize norms, values and skills that can determine variable propensities for mobility among different individuals and, in the case of the individual, in the various stages of their life (Vincent-Geslin and Ravalet 2015). This perspective refers to the dimension of *mobility biographies* (Axhausen 2008; Scheiner and Holz-Rau 2013) as the set of events or circumstances such as the transition from one age to another (Rau and Scheiner 2020), access to economic resources or specific means of transport (e.g., obtaining a driving licence), and changes in place of residence and work that influence our mobilities and their temporality, resulting in moments of significant movement followed by more or less long periods of relative stability (Lanzendorf 2003). Conceiving the relationship between mobility and immobility in relational terms thus means recognizing that all individuals directly experience immobility throughout their days and lives. At the same time, another consequence of this reflection is that we question whether mobility and immobility take absolute forms: just as systems of 'hypermobility' (Urry 2002) require moments of immobility and anchorage, so forms of immobility become meaningful when considered in light of the mobilities of which they represent an interruption, or which they replace with different, minimal, unobservable or unaccustomed mobility practices that we mistake for immobility. Thus, the differences in these movements constitute what Adey (2006) defines as relative immobilities that must be read as a function of individuals' possibilities and capability to move and stop when, how and where they want to during their life experiences.

While the relational perspective allows shedding light on the continuous intermittence between moments of mobility and immobility, the interpretation of the relation in differential terms helps to explore the possible individual or contextual conditions that influence the breath-like rhythm regulating such intermittence (Cresswell 2010 in Motte Baumvol et al. 2015), facilitating and supporting the movement of some while, on the other hand (and simultaneously), inhibiting, limiting or preventing that of others (Massey et al. 1993). Mobility is still conceived as a fundamental activity in which the actor is not a neutral agent of the movement but represents an active protagonist based on the mobility capital at their disposal that is inscribed, in turn, within a global spatial capital (Lussault 2005). The concept of *motility* (Kaufmann et al. 2004) encompasses the elements that determine the mobility capital of each individual. These are *access*, as the range of possible mobilities that can be acted considering spatial and temporal constraints; *competence*, as the set of skills and knowledge that support and allow movement; *appropriation*, which represents the actual materialization of the mobility plans of an individual (Kaufmann 2002; Kaufmann et al. 2004). Therefore, the availability of a mobility capital is related to the characteristics and possibilities of the individual and the territorial, social and cultural contexts of belonging (Cresswell 2010), which explains why the distribution of such capital is uneven across different social groups. The forms of mobility and immobility of individuals are considered as essentially influenced by the unequal power relations that structure society along ethnic, gender, economic status and educational lines (Sheller 2018a), having the potential to both physically and socially immobilize people alongside more fluid social arrangements (Rau and Vega 2012). At the same time, the reiteration of such practices contributes to the constant (re)production of factors of inequity, making mobility a pressing phenomenon of social stratification (Bauman 1998). Mobilities are and remain an essential element in many aspects of an individual's life, especially in contemporary social configurations in which networks of inter-individual relationships become increasingly complex and spatially dispersed. More extensive and erratic mobility is therefore required to access and, at the same time, maintain such access to these networks that provide interaction and proximity with others (Urry 2005). For this reason, lack of mobility is the real problem for many social groups (Urry 2002; Farrington 2007) who are excluded from essential networks and opportunities. Delving into the knowledge of said lack, which is at the origin of the differential expressed by the concept of relative immobility, becomes essential to identify the nature of the constraints that can determine this differential, the socio-spatial impacts that it produces, and the individuals or social groups that suffer from it. A discourse on immobility thus takes on a solid value in understanding the social, spatial and infrastructural inequities that are expressed through—and, in turn, produce—differential mobility practices, laying the groundwork for subsequent explorations aimed at establishing of what some authors identify as 'Mobility Justice' (Sheller 2018b).

Regardless, it is nonetheless significant to reaffirm that mobility and immobility are both differential and relational and that the experience of different rhythms and practices of mobility can be traced to reasons and take on very different meanings from individual to individual.

2.3 Immobility and Social Inclusion/Exclusion

Since mobility represents a primary component of our human experience, being conceived by Ascher (2005) as a right that, to at least a minimal extent, societies should guarantee to individuals to allow them to meet their desires and expectations, it is intuitive to attribute a negative valence to immobility (Motte-Baumvol et al. 2015). It is even more logical to consider such condition as the effect of forms of coercion that oppose the exercise of vital practices to the individual, ultimately producing social exclusion.

The concept of social exclusion related to mobility is significantly discussed in scientific literature. Here we will use the well-known definition proposed by Preston and Rajé (2007) inspired by the seminal work of the Social Exclusion Unit (2003), which considers exclusion as a process induced by the existence of barriers that make it difficult or impossible to participate in the activities of the societies to which one belongs. The same authors emphasize how this constrained-based process has intense spatial manifestations, while Schwanen et al. (2015) affirm its multidimensionality—that is, the fact that it can simultaneously depend on various factors not only attributable to the economic conditions of the individual, but also to social, geographical and cultural aspects—, dynamism over time and its multiscalar effect, involving different social spheres at the same time.

Taking up the definition of motility proposed by Kaufmann et al. (2004) and introduced in the previous section, the concept of mobility-related social exclusion can be associated with the impossibility of disposing or using a capital of mobility (motility), a spendable capability to access all those states of being and doing, defined as functionings, that a person has reason to value (Sen 1990). If being mobile in societies built around the assumption of high mobility is necessary to gain access to formal and informal networks of work, leisure and friendship (Kenyon et al. 2002), forms of insufficient mobility due to the existence of barriers inhibiting movement can thus be a consequence of a lack of mobility capital and be at least partly at the basis of forms of social exclusion (Cass et al. 2005).

Barriers to movement can take on many forms and behave differently depending on the type of mobility they contribute to immobilizing. Resuming to the concept of motility, all those person-based and context-based circumstances that influence its three components (namely access, competence and appropriation) and that produce an impact on our mobility behaviors can be considered as barriers to mobility. The study of the factors that influence travel behavior has produced conceptualizations trespassing urban and transport studies, sociology and psychology aimed at investigating not only upon which more or less rational logic people move or do not move in a certain way, but also what the regulatory, spatial and economical devices able to change these behaviors can be (Næss 2005; Martens 2006; te Brömmelstroet 2014).

Barriers that can affect the dimension of *access*, namely the individual's potential mobilities aimed at seizing spatial opportunities and transforming them into something useful (Martens et al. 2012) relate to the unavailability of transportation, communication devices and skills, and information through which to organize and accomplish one's movements. Moreover, barriers to access are also constituted by the lack of meaningful and desired opportunities that can be reached when time and cost restrict one's activity patterns (Hägerstrand 1975; Kaufmann et al. 2004; Kaufmann and Viry 2015; Lyons and Davidson 2016). The dimension of physical access thus becomes a key to qualifying the conditions of social inclusion and exclusion related to mobility practices and immobility of individuals. As elaborated in the following section, a focus on the concept of accessibility to places and activities conceived as a potential for participation (Martens 2017) allows unveiling the lack of suitable transportation or accessible activities affecting individuals or geographical areas that constitute limits to their social inclusion (Kamruzzaman et al. 2016).

From this point of view, relative immobility can emerge because of the absence of opportunities attainable through one's practices of spatial mobility resulting in a consequent limitation in the frequency and intensity of such practices due to high travel expenditure, time and space restrictions, and long travel times (Oviedo et al. 2019).

Linked to the dimension of access is that of *competence*, which considers the entire set of skills, knowledge, aspirations and abilities related to the person's socio-economic and demographic conditions that are required to overcome spatial—temporal constraints and access the resources and activities we need and desire (Schwanen et al. 2015), as well as the possible effects of rules, laws, behavioral codes and preferences that may impact how and how much we move, besides the emotions we associate to movements (Cresswell 2010). Reasons for (not) moving are the result of the perception or the actual possibility of physically accessing spatially distributed activities and opportunities, but also depend on the needs, attitudes, habits and goals that each individual associates with movement and that are shaped by the adherence to social, legal, and symbolic norms that we refer to while carrying out our mobility practices (Dijst et al. 2013). Social, political and cultural contextual factors thus add up to spatial and temporal frictions that can induce or limit an individual's mobilities, as in the case of the so-called coupling constraints, namely those normative and social regulations that require the individual to physically participate in certain activities based on the grounds of production, consumption and social contact (Hagerstrand 1975). A similar effect, in terms of limitation of movement, is driven by moral and ethical factors pushing the individual to make choices and actions that are appropriate to one's values or that are approved by others (de Groot and Steg 2008), as in the case of reducing travel and its frequency to limit one's ecological footprint. Political and social factors can also be reflected in the construction of physical-spatial, legal and perceptual barriers (Harker 2009; Motte Baumvol and Nassi 2012). An example of such barriers can be found in the complex systems of control developed to coercively and differentially counter and regulate the physical mobilities of certain ethnic or social groups such as refugees, migrants and other foreigners in the context of what Turner (2007) identifies as a regime of global immobility founded on complex systems of surveillance and control. Although in many cases these barriers have a specific physical connotation made of fences, walls, technological infrastructures of screening, monitoring and registration capable of regulating the physical access of individuals and things, they have a legal connotation since they are legitimated and guaranteed by rules and laws of states. The objective of these global security systems is to regulate spaces and flows of people, goods and services according to a differential logic balanced between global openness and maintenance of sovereignty over their borders in which the immobility of some becomes an integral part of the maintenance of privileges (also expressed as seamless mobility) of others (Cresswell 2010). Furthermore, the action of barriers is reflected on the perceptual plane of those who are immobilized, contributing to a different temporal and spatial understanding of their possibilities of access that can give rise to new forms of different, desperate mobility (Martin 2011) as a reaction to the action exerted by legal constraints that produce spatial and temporal restrictions to which they are subjected (Harker 2009).

Consequently, the highly personal ways in which we *appropriate* the mobility resource, that is, how we act with respect to the possibilities of access, availability of skills, and capabilities, whether these are individual or contextual, actual or perceived, can generate relative immobilities with different causes and effects depending on the individuals that experience them and the geographical, cultural and political realities in which they live. As seen, these immobilities may be the product of conditions of social exclusion and may, in turn, contribute to reproducing them.

However, it is still possible to question a bi-univocal relationship between conditions of immobility and social exclusion. Different people may be both enabled and constrained in relation to mobility practices, and the resulting immobility may describe a different set of more or less desirable experiences away from a clear-cut logic differentiating them (Adey 2006; Cresswell 2012; Straughan et al. 2020). We thus return to the idea of immobility as an essentially differential and relational phenomenon, influenced by social inequalities (Massey et al. 1993), in which the distinction between being socially included or excluded is not only connected to how extensively and frequently we move, as a totalizing conceptual approach to mobility would suggest, but also to the degree of voluntariness with which we choose or not to do so.

The possibility to self-determine the personal bodily mobility, choosing when and with what specific mobility networks to move, is an important form of power in contemporary hyper-mobile societies, and it is assumed that freedom of choice in one's mobility represents the affordance of the right to move at will (Martin 2011). The consequence is the emergent conception that greater spatial mobility reflects a higher possibility of participation in social networks of increasing prestige. Nevertheless, the same can be also said for all those forms of more or less intense relative immobility that we can experience (Murphie 2011) when they represent the product of choice made by the individual that does not affect their perception of inclusion and participation. In this case, less mobility does not mean less power for self-determination and inclusion, since it is the possibility and the capability to choose one's state of motion or stillness that becomes in its way a form of power.

Numerous instances of this perspective are documented in the scientific literature, beginning with Cresswell's (2005) historical analysis. The author asserts that the association between mobility and social privilege, foundational to the concept of the kinetic elite, is a relatively recent construct. The meanings and practices of mobility are fundamentally ideological; historically, mobility has not always been synonymous with freedom, citizenship, and inclusion. The literature provides examples such as the medieval wanderer and gypsy, uprooted figures living on society's margins and viewed suspiciously by the inhabitants of fortified cities, or the vagabond and the straggler (Le Breton 2005), whose errant mobilities Bauman contrasts with the orderly, legitimized movements of the tourist (Bauman 1998). Further nuances in the nexus between mobility and immobility are highlighted by Jocoy and Del Casino (2010). Their study of homeless individuals' mobility in the U.S. reveals that immobility can represent a privilege. Homeless people, often coerced into constant movement between shelters, are precluded from resting in public spaces due to negative societal perceptions and legal constraints that view them as a threat. Paradoxically, their only moment of stasis occurs when using the bus as a shelter, where a ticket allows them to remain in place, equating, at that moment, their nomadic condition to that of other passengers.

Other traditionally immobile and inactive social groups, such as the unemployed, are frequently forced into erratic movement in search of work in a society that stigmatizes those not engaged in the economic system. Here, stillness is primarily interpreted as missed productivity (Bissell and Fuller 2010; Marston et al. 2019).

Still looking through a relational and differential perspective, just as some individuals' mobilities depend on others' immobilities, certain forms of immobility may also rely on the mobilities of others. A compelling example of this interdependence is observed during the lockdowns imposed during the Covid-19 pandemic. Individuals who worked remotely, especially those in high-value-added sectors, experienced a relatively privileged condition, safe from contagion and forced contact with strangers, and were able to reassemble their means of work, health, reproduction and ultimately survival within their own homes (Adey et al. 2021). Instead, on the outside, essential service workers, often poorly protected, unrecognized and belonging to minorities (European Institute for Gender Equality 2021), have continued to move around deserted cities offering vital support to the immobile survival of others. In this way, rather than spatial mobilities, immobilities became the luxury of the economically privileged (Marston et al. 2019).

Additional examples of these interdependent relationships can be found in the complex mobility chains associated with work-life balance, especially when caring responsibilities require that some family members provide support to others who, for various reasons, cannot act their mobilities independently from others. In other circumstances, the immobilization of some family members is instead the result of the mobilization of others, as occurs in the cases of the wives of long-distance workers, who find themselves in a condition of perpetual waiting and stuckness that produces both forms of spatial and emotional immobility exacerbated by the high mobility of their partners (Straughan et al. 2020).

Finally, it is also interesting to consider the possibility that relative immobility could be a state of being intentionally adopted by the individual. Conradson (2011), for example, sees immobility as a way to escape the hurried rhythms of life that many perceive as a common and founding trait of contemporaneity. According to Bissell and Fuller (2010), a quest for stillness—and, by extension, slowness and immobility—can be seen as a response to "*the desire to bow out of the relentless and often-ugly rat race of capital accumulation for more prolonged durations* [that] *is immortalized in many channels of popular culture, where idleness and rest are among the chief characteristics*" (p. 5). Phenomena such as the return to the countryside, marked by the advent of new figures such as the amenity-led migrant (Moss 2006) or the *mountaineers by choice* (Dematteis 2018), and more generally of all those forms of deceleration that put into practice an idea of life favoring relationships and mobility of proximity, exemplify how immobility can be a sought-after and desired state. Indeed, following the interpretation of van Wee (2021), immobility can be conceived, like mobility and accessibility, as a positional good to which a certain status is conferred and associated with specific positive perceptions that produce voluntary immobile behaviors.

In paraphrasing Le Breton (2005), we can therefore argue that there is no simple and immediate relationship between mobility, immobility and an individual's social position. What is commonly accepted is that mobility participates in increasingly defining differentials between those who have access to composite and varied worlds and those excluded from them. However, the essential point is that access to these inclusionary worlds does not have to be only attained through high mobilities, but can also happen through forms of relative immobility, provided that they are *reversible, or chosen,* and not *constrained.* As we have seen, these forms can in part depend on the mobilities of others but can also result from conditions of proximity and the possibility of physically and virtually access desired and necessary networks and opportunities.

2.4 Spatial Implications of Immobility and Accessibility by Proximity

Physical proximity is to be considered one of the reasons that push individuals to move. According to Urry (2002), physical displacement is fundamental to allow for forms of co-present interaction and face-to-face encounters that are necessary to maintain and strengthen ties within increasingly fragmented and spatially dispersed social networks (Urry 2002; Elliot and Urry 2010; Larsen 2014). This capacity consolidates individual social capital, which, according to Bourdieu (1986), depends on the extent of the network of contacts and resources that an individual can actually or potentially access, thereby facilitating access to various types of economic, cultural, and symbolic capital. Mobility thus becomes a resource that enables encounters through movement acting as a multiplier of opportunities for participation, involvement and inclusion. Because of its very characteristics, the drive towards physical proximity is felt to be compulsorily appropriate or desirable by people; this is the reason why a lack of mobility will represent for some a limit to their participation, which they will try to overcome by resorting, as soon as possible, to greater mobility (Urry 2002).

The combined desire and need for physical proximity to other members of one's social network, alongside technological innovations that allowed moving more quickly and cheaply on large geographical scales such as cars, has determined, especially in the last seventy years, massive growth in the distances covered and the speeds reached by individuals (Banister 2011). These are also compounded by the possibility that the time spent traveling on a daily basis may increasingly exceed the maximum threshold tolerable by individuals according to Marchetti's constant (Marchetti 1994), as it seems to be the case in cities developed in the age of the automobile, which, due to their size, prevalent mono functionality and sprawling, can become dysfunctional (Newman and Kenworthy 2015) or in the case where the dispersion of activities that characterizes living contexts out of the main urban core drives people to travel for an extended time to reach the first available spatial resources with a negative impact on their wellbeing.

The increase in mobility, which is expected to keep growing in the following years worldwide (Sustainable Mobility for All 2017), does not seem to have been reversed by the effect of virtual information and communication systems (ICT). As van Wee et al. (2013) point out, much of the early optimism about substitution has been reduced over time, with a still strong preference for face-to-face interactions. On the contrary, ICTs have proved useful supports in expanding our mobilities, often becoming complementary triggers to movement and relevant factors of change and reorganization of our daily activity patterns (Kenyon et al. 2002; Lyons 2014; Lyons and Davidson 2016; Vecchio and Tricarico 2019). ICTs have also contributed to the acceleration of economic, social and cultural processes and practices that some authors identify as a *characterizing aspect of our modernity* (Rosa 2009, p. 78). According to some authors, this acceleration, which is also expressed through the establishment of increasingly complex and costly global hypermobility systems, could paradoxically lead to a kind of loss of the individual exposed to increasingly virtual reality that proves challenging to be perceived and understood. The result of this process would be the emergence, as a reaction, of new forms of bodily and political inertia (Virilio 1999) as conditions of immobility and powerlessness of the body and spirit in an increasingly unreal world in which virtual spaces dissolve the scale of human environment (Gane 2006).

Although Virilio's radical view from the late 1990s has been partly contradicted by subsequent events—at least regarding the frequency and intensity of physical mobility—it has gained renewed interest in the Covid-19 era. The conditions imposed by the Pandemic seem to have, at least for some social groups, created the basis for generating potential "light" forms of polar inertia (Virilio 1999) in which virtual technologies have partially replaced mobility and encounter, especially for those movements related to activities that can be performed in digital format. As interesting as such a perspective may be in imagining future scenarios, a clearer picture of actual mobility behaviors post-pandemic is now emerging after some years has passed from the lifting of the most rigid "mobility saving" solutions imposed during the lockdowns and their long-lasting impacts on our mobility and immobility behaviors can be now better appreciated. Evidence suggests that daily mobility rates in Europe have not yet returned to pre-pandemic levels (Christidis et al. 2022), possibly due to other factors impacting economy such as the Russian invasion of Ukraine. However, this reduction in travel is different between modal choices, as seen in Italy (ISFORT 2023), where public transport usage has significantly decreased and, despite gradual recovery, still remains below pre-Covid levels. This phenomenon is attributed to health concerns leading to behavioral shifts toward private vehicles, driven by psychological factors (Aaditya and Rahul 2021). In contrast, Car travel has consistently increased, while active mobility (walking and cycling) saw a significant rise during the pandemic but stabilized post-restrictions. Interestingly, proximity mobility (namely all those trips shorter than 10 km) has increased post-pandemic in Italy, suggesting that limitations on commuting for work and study imposed during the pandemic have led to a stabilization of the number of relatively immobile labor force working from remote. Indeed, while global remote working numbers are lower than during lockdowns, they remain higher and more stable than pre-pandemic, with many cities experiencing significantly fewer people returning to offices in 2022 (Brouwer and Mariotti 2023). Factors such as commute costs, care responsibilities, and lifestyle attitudes contribute to the preference for remote work (Silver 2023). This trend is confirmed in Europe, where commute distances significantly influence the choice to work from home (Dias da Silva et al. 2023), demonstrating that savings in travel time and monetary resources are fundamental in shaping behaviors and preferences. Although remote working numbers were negligible before 2020, it is likely that demand for remote work will remain substantially higher than pre-pandemic levels (ibidem).

Other effects of the pandemic on daily mobility have been found in the field of remote learning. Used in many countries around the world as a solution to limit the possibility of contagion during the pandemic, remote learning has highlighted various problems and inequalities linked to the quality of children's learning associated with factors such as socioeconomic status, access to technology, the quality of the learning resources and environment, the teachers' feedback (Cortés-Albornoz et al. 2023). While the long-term impacts of lockdowns on learning are still under study, remote learning has continued post-lockdown and is today seen, at certain conditions, as a promising way to deliver education through innovative formats. This is particularly true for higher education since remote learning offers higher flexibility and continuous availability of learning materials, thereby increasing higher education access for previously excluded populations (Santos 2022).

Also, the pandemic has underscored the importance of spatial proximity and service distribution in urban planning, during a period when severe mobility restrictions highlighted the necessity of accessing essential daily services within short distances and times. The attention paid to the design of public spaces and their degree of accessibility, quality, and comfort has also increased, recognizing their fundamental role in personal well-being, especially when a significant portion of the global population was confined to indoor spaces. Consequently, the pandemic has imposed a change of perspective for urban policies, accelerating or confirming already active urban policies aimed at enhancing proximity and active, sustainable mobility. This shift is not only deemed necessary during the emergency phase but is also increasingly viewed as a promising avenue toward more sustainable and inclusive future cities (Lanza and Pucci 2022). Nevertheless, even before the outbreak of the Pandemic, there were numerous calls, in scientific literature and increasingly also in urban policy, to reconsider the value of forms of counter-representation of the hypermobile paradigm and the positive social and environmental implications on promoting a low mobility, or relative immobile, society (Vannini 2014; Holden et al. 2020). This last concept, the origin of which can be traced to Banister's seminal paper (2008), is based on the idea that mobility represents a high valued capital necessary for our lives, but that frequent and long-distance forms of travel produce negative externalities such as pollution, greenhouse gases, congestion and occupation of city living spaces, and intense frictions in people's lives and psychologies (Bertolini 2012). Moving is a necessity, but, as many of us are dependent on high mobilities to conduct our lives, it is a necessity that is often met in environmentally and socially unsustainable ways (Banister 2008; Cohen and Gössling 2015; Ferreira et al. 2017; Mattioli et al. 2017).

The idea proposed by Banister (2008) with the sustainable mobility paradigm is that mobility behaviors in contemporary societies are, at least in part, influenced by a combination of social, spatial and cultural factors that characterize the territorial contexts in which the person resides and move. In this way, the basic social need for physical proximity and face-to-face encounters underlying our desire for mobility becomes unsustainable if it translates into practices of high mobility that are time and resource consuming (Pyrialakou et al. 2016) and mostly supported by private and individual means of transport. Already in 2000, Putnam pointed out how specific spatial configurations and the behaviors they were shaping, such as high dependency on high frequency and long-distance car-based mobility, were at the origin of forms of decline in local social capital and disconnection from one's neighbors. Thus, the maintenance of forms of social capital on dispersed networks creates a potential distancing from closer social and spatial relational opportunities and produces, at the same time, serious negative externalities for the community.

The principle of a low mobility, or relative immobile, society is grounded in the assumptions of the city's spatial, functional, and infrastructural organization, providing a blend of land use and transport measures aimed at reducing the need for travel while increasing accessibility, here understood as the ability to participate in valued spatially-distributed activities influenced by a set of individual and contextual characteristics (Geurs and Wee 2004; van Wee and Geurs 2011). In this context, accessibility encompasses both spatial or transport-related constraints (contextual characteristics) and personal capabilities, possibilities, and preferences (individual characteristics), which affect social and activity participation. However, accessibility can be achieved through various ways. According to Levine et al. (2019), it can occur through mobility, involving physical movement supported by transportation systems to reach spatially distributed activities; through proximity, which entails physical closeness to specific activities such as service facilities; and through connectivity, which involves the delivery of goods and services to one's location. Additionally, accessibility can arise from what Manzini (2021) describes as relational proximity, defined as the result of intense interpersonal connections among individuals sharing specific physical and social spaces, where opportunities for collaboration and mutual support can emerge. Indeed, what this last perspective proposes is that accessibility needs may be fulfilled through community-based social relations, collaboration, and exchange, rather than solely through physical or virtual means.

Among the different ways in which accessibility can be achieved, two are particularly relevant to this research and consistent with the principle of a low mobility society in a post-pandemic era, namely all forms of access based on proximity and connectivity, which can reduce our need for extensive and long-distance travel in daily life, thus creating the ideal conditions for the development of what has been previously termed reversible immobility. This reflection has significant practical implications, as it can reorient transportation and land use planning from merely improving transport systems to favor seamless mobility towards increasing the availability of services, activities, and social contacts in proximity. These planning principles towards proximity and low mobility are encapsulated in the theoretical and operational concept of accessibility by proximity, introduced in the literature (Pucci et al. 2022) and applied in several cities worldwide, even before the Covid-19 pandemic, but strongly accelerated after its outbreak. Examples include the 20 min neighborhoods experienced in Portland and Melbourne, the more recent 45 min cities with 20 min towns in Singapore, and also the well-known "ville du quart d'heure" proposed in Paris. This concept aims to create the conditions, through planning policies and measures, to make cities more inclusive and sustainable by reducing the spatial and temporal intensity of our displacements by providing accessible services and opportunities in physical proximity, stimulating social interaction in high quality pedestrian-oriented public spaces where community ties can be strenghtened (Handy et al. 2002; Ferreira et al. 2003, 2007, 2012) and potentially leading to the development of forms of relational proximity according to the reflections proposed by Manzini (2021). Also, accessibility by proximity inspired policies promote sustainable forms of active, collective, and shared mobility and digital access to goods and services.

If in Urry's perspective physical proximity represents one of the purposes of (hyper)mobility, it becomes both a design and normative tool under the concept of accessibility by proximity which can conversely induce forms of relative 'immobility of proximity.' Such behaviors could be chosen by individuals precisely because the contexts of everyday life, through their spatial configuration and the access they provide to social and digital networks, goods, and services, enable them to obtain their necessary resources and multiply opportunities for high- and low-intensity physical contact with others (Gehl 2011). Furthermore, by emphasizing the value of accessibility, proximity, and connectivity to people and things, planners and policymakers can explore issues related to behavior, culture, use, and design of living spaces, and how increased sensitivity to these issues can innovate planning practices (Banister 2011). In brief, proximity to people and things should be not only conceived as something guaranteed and sustained by high mobility practices but can be intended as a factor that, if promoted through effective planning mesaures, increase the livability of urban space and local social interactions, thereby replacing the need for high mobility.

As interesting as it is, the low mobility model inspired by the concept of accessibility by proximity and its various policy translations comes with several open issues, particularly regarding the translatability of this ideal type into concrete terms. Among these, scientific literature recognizes the need to differentiate the policy measures inspired by the concept, considering how travel is a differently experienced activity in the varied urban contexts and related settlement conditions existing globally. Moreover, a push towards a low mobility society is contrasted by the strength of culturally embedded habits acting towards the maintenance of certain established travel behaviors, especially those requiring high mobility and heavy use of polluting means such as cars and planes that are attributed an essential social and economic significance (Cresswell 2010; Dijst et al. 2013). Similarly, planning for alternative models of urban development questioning the status quo can generate political contrasts and conflict between divergent interests (Holden et al. 2020) and even lead to unexpected and contradictory effects. An example of such a negative outcome is described by the hypothesis of opportunity (Næss 2005), stating that the monetary savings allowed by practicing daily mobility over shorter distances could be reused to increase the length and frequency of other, less systematic movements. Also critical in this discussion is the hypothesis that establishing a low mobility society model could jeopardize accessibility to some social groups dependent on transport needs that have become essential in modern societies. Therefore, attempts to reduce the need for mobility may generate inequality and disadvantage for certain groups whose social inclusion mainly depends on high mobilities or the use of means they can afford but that are considered no longer sustainable (Mattioli et al. 2017; Holden et al. 2020). The attempt to ensure inclusion through greater accessibility rather than through greater mobility (Preston and Rajé 2007; Kamruzzaman et al. 2016) must therefore account for what benefits and costs are induced by policies inspired by the low mobility model and how they are distributed over members of a society (Schwanen et al. 2015; Benenson et al. 2017; Martens 2017). Indeed, the risk of displacement of certain social groups due to rising housing costs from areas of high concentration of functions and accessibility could challenge the effectiveness of these policies by expelling inhabitants from the denser areas towards longer commutes, ultimately generating social and territorial imbalances (Cervero 1989; Levinson 1998).

Regardless of these limitations, a low-mobility society is still seen as a possible solution for a perceived resilience problem of contemporary mobility systems. Indeed, some authors, including Ferreira et al. (2012, 2017) argue that the increasing social, technological and managerial complexity of these systems, on which the functioning of the global economy depends, is augmenting their vulnerability. According to the authors, it is necessary to explore the positive implications of a shift from the development paradigm based on globalism and high mobility towards models inspired by stillness, slowness and localism. In line with Banister's proposals, proximity-based accessibility is attributed the driving role in configuring a society able to dispose of and multiply new forms of social and spatial capital without depending on high mobilities. Reversing the concept of motility proposed by Kaufmann et al. (2002), the authors introduce the notion of *immotility*: immobility, if reversible, is thus conceived as an actual form of capital in societies in which low levels of mobility are not perceived as the result of a constraint or the action of barriers, but rather as a positive factor of relationship and proximity with one's immediate physical and social surroundings. In other words, relative immobility, like mobility, can be a mean to the end of social interaction and participation when the prerequisite of accessibility by proximity is present.

The review confirms, as synthesized in Fig. 2.1, that there is a lively academic debate, strongly influenced by the outcomes of the Covid-19 pandemic, concerning the implications of models of reduced mobility and how forms of reversible immobility can be induced by spatial and social configurations based on the principles of spatial proximity and accessibility. Hence, the field is cleared of the misconception that immobility can only be the product of forms of exclusion by proposing instead the idea that immobility may be, per se, a form of social and spatial capital.

2.5 Defining and Analysing Immobility Through a Theoretical and Operative Framework

Based on the literature review, and bringing back the issue back to the disciplinary field of urban and transport studies within which this work is placed, a definition of relative immobilities is proposed that can guide their exploration, dealing with the complexities and ambiguities that characterize the different ways in which these conditions are manifested and experienced. Relative immobility, in this book, is defined as the result of a mobility differential between individuals emerging as a reduced temporal and spatial extent of daily movements. Depending on the levels of accessibility by proximity related to the functional characteristics of the context of life and the individual possibility of meeting their needs and expectations, such a differential may (in the case of constrained immobility) or may not (in the case of reversible immobility) be traced back to a potentially limited social participation and exclusion. It follows that the concept of immobility proposed here mainly concerns the differential dimension of immobility by focusing on the sphere of the individual's daily practices—and, for extension, of the possibility for social participation—that are expressed as physical movements in time and space, and how individual and socio-spatial contextual factors influencing accessibility by proximity can contribute to limit or favor them. Thus, a simplification is made, placing in the background, but not forgetting, other highly relevant issues concerning the relative dimension and intermittence between mobilities and immobilities and how they may change over

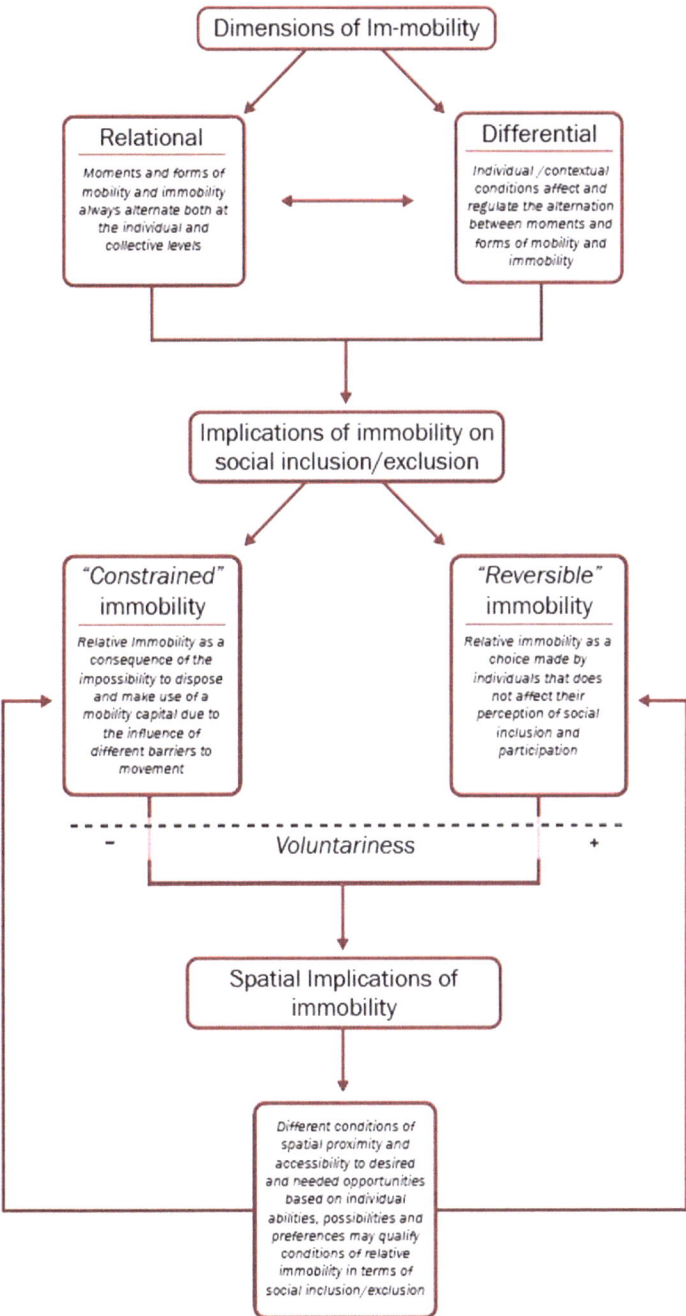

Fig. 2.1 Synthetic conceptual scheme

an individual's (daily) life. Also, this framework mainly focuses on physical accessibility by proximity rather than digital connectivity even if, as clarified in Sect. 2.4, indirect accessibility allowed by connectivity could be considered a central contributor to accessibility, as clearly illustrated by Levine et al. (2019). This methodological choice is justified by the fact that this research work is inscribed within the disciplinary field of urban planning and adopts a lens that focuses on the spatial dimension of immobility. Therefore, it is deemed important to concentrate primarily on the component of physical proximity rather than connectivity, which could be a relevant object for future research and will, in any case, be extensively mentioned in subsequent chapters of the book, albeit without proposing dedicated measurement methodologies to assess the level of access granted by digital networks. Furthermore, following the proposed definition, the difference between relative spatial mobilities and immobilities is intended both a matter of time—related to frequency and duration of movements—and space covered through movement. Clearly, an individual could devote a high amount of time to travel by moving frequently in a very limited space and, viceversa, move very unfrequently but on high distances. These ambiguities can make it particularly complicated to sharpen the already blurred boundary between being mobile or immobile. This requires the development of methodologies and tools specifically designed to investigate and explore these conditions, as proposed in the subsequent chapters of the book.

The proposed definition linking mobility, immobility, and accessibility by proximity lies at the base of the conceptual framework (Fig. 2.2) that orients the development of planning-related reflections and analytical methodologies to better understand the socio-spatial implications of immobility and to achieve the objectives of (i) highlighting how, in varying socio-territorial conditions, relative forms of immobility can represent both a form of potential deprivation and a reversible choice with positive social, economic, and environmental impacts; (ii) as a consequence of the previous point, to directing more specific and context-based analyses for the design and implementation of public policies of social equity improvement with which to reverse, if necessary, situations of immobility-driven relative disadvantage or deprivation that are difficult to detect through traditional mobility-centered tools and methods; (iii) to promoting an 'emancipation' of the concept of immobility by questioning the cultural nexus that associates forms of spatial immobility with a status of social immobility and exclusion, revealing, through the empirical methodologies that can be developed from this framework, a more varied and complex range of cases. From this point of view, if forms of immobility associated with good levels of accessibility to opportunities may not be attributable to a problem, forms of hypermobility forced by the absence of proximity opportunities may instead represent more of a burden than a capital available to individuals; (iv) to advancing the idea, in line with the previous point, that the promotion of reversible forms of immobility can even be assumed as an objective of urban policies aimed at the (re)organization of land use and transport systems, moving away from a mobility-centered perspective that, as seen, can in turn generate further conditions of imbalance, social exclusion and environmental unsustainability.

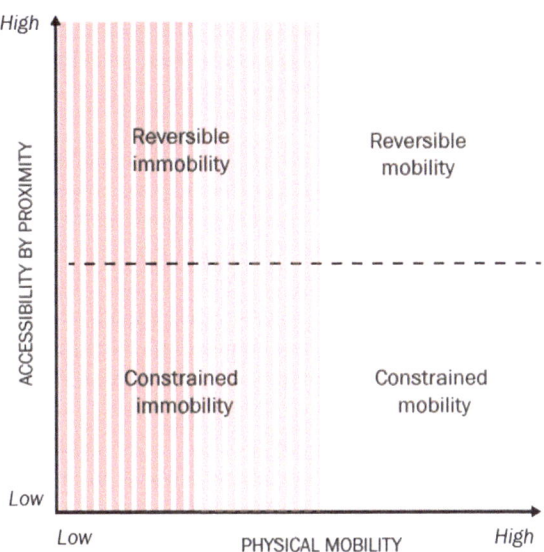

Fig. 2.2 Conceptual framework linking reversible and constrained forms of immobility

Analyzing the framework in detail, four main profiles emerge and can be systematized. When low levels of daily mobility and travel in time and space experienced by individuals in certain geographical areas are coupled with poor accessibility by proximity, a shortage of mobility options could be in place (Kenyon et al. 2002). This condition is referred to as constrained immobility because it is associated with a limited potential for social participation expressed by relatively low levels of accessibility. Therefore, those who live in this condition may suffer from insufficient mobility, which is not compensated by good accessibility, potentially excluding the individual from social participation.

A different situation characterizes those who live in conditions of good accessibility to proximate necessary and desired opportunities with low levels of physical mobility and travel. These profiles express a condition defined as reversible immobility, because it does not compromise, at a theoretical level, the potential social inclusion and participation of the individual, determining, as we have seen, a series of potentially positive collateral effects in social and environmental terms.

The other profiles that emerge from the relationship between accessibility and mobility concern conditions characterized by high levels of mobility, which have been extensively studied in scientific literature and that, as argued in the previous chapter, may in turn refer to conditions of potential disadvantage and exclusion.

On the one hand, extended mobility in time and space coupled with high physical accessibility by proximity can describe the multiplication of opportunities for movement and participation generated by the extension of one's networks over highly infrastructured and dynamic geographic areas. This condition, defined as reversible mobility, indicates a situation in which the individual's behavior leads, at least in part and for some activities, to prefer more remote options than those available nearby. In this condition, it is possible to observe one of the types of access proposed by

Levine et al. (2019), namely accessibility granted by mobility. Such behavior could be redirected towards less intensive or, at least, more sustainable forms of travel without theoretically compromising the individual's possibilities for basic activity participation.

On the other hand, forms of high mobility coupled with low accessibility by proximity could signal a capacity offered by the context and by individual predispositions to reach remote but accessible opportunities; however, at the same time, they can be interpreted as a signal that long-range mobility, (expensive in terms of time, economic and environmental resources) are potentially required to reach desired destinations. Therefore, this profile of constrained mobility could be particularly interesting to deepen because it identifies contexts in which immobility could be not so much of a problem to be confronted (as in the case of conditions of immobility coupled with low accessibility) and rather become a possible target for immobility-enabling integrated land use and transport policies to be implemented in these contexts.

It is worth noting that this framework is proposed in the awareness that the conditions of accessibility to a range of activities, even those relevant to meet some basic needs of daily life, cannot exhaustively explain the totality and complexity of the patterns of immobility expressed by the population settled within a territory. However, the approach proposed in this book does not aim to identify causal links between mobility and accessibility, but rather to identify those situations in which being more or less mobile can affect the level of potential activity participation of an individual according to the availability and accessibility of opportunities in physical proximity.

Beyond this general point, it is important to note that the framework also introduces other highly relevant issues derived from the definition of immobility provided in the chapter that should be object of further scrutiny and research. These issues concern various aspects of transitioning from a theoretical conception of immobility and its relationship with accessibility by proximity to translating these concepts into operational terms to generate value for planning policy. Four particularly relevant issues warrant more detailed focus and will be addressed accordingly in the following chapters of the book: firstly, two fundamental questions emerge related to (i) how to measure and quantify immobility and (ii) how to measure accessibility by proximity, that is evaluating the spatial conditions enabling chosen immobility. Indeed, the framework compares two dimensions that can be measured through very different methodologies and techniques. This applies to the quantification of realized immobility and the existing mobility differentials between different individuals and social groups, as well as the quantification of accessibility levels to specific spatially distributed activities. Equally challenging is the possibility of defining gradients to establish when daily mobility practices could be labelled as 'low intensity' both in time and space, giving rise to forms of relative immobility. The same is also proposed for accessibility and its measured levels. A possible answer to the problematic issues arising when defining such gradients is based on the idea that levels of immobility and accessibility to opportunities are to be analyzed in relative and contingent terms according to a scheme that can be alternatively or simultaneously context-based or person-based. In particular, the framework should be applied, as previously suggested, to identify differences in the frequency and geographic extent of daily movements expressed by single inhabitants or groups living within a defined territory and relate them to the different levels of accessibility to opportunities in the same context, producing diverse geographies of immobility characterizing the different geographic and social environments that make up the study area. Relativizing the levels of mobility and accessibility by referring to a territory and its existing social and spatial conditions is a way to detect and quantify the extent of potential inequities and differentials in mobility and access to opportunities useful to promote equity through planning and transport policy in a more context-sensitive way. A focus on the measurement and analysis of these conditions is thus essential to quantify these phenomena, describe them, interpret them, and address any problems related to their presence and characteristics.

The two preceding issues particularly concern questions related to the analysis and measurement of levels of immobility and accessibility by proximity as indispensable activities for their understanding at a territorial level. However, the literature review shows that immobility is a phenomenon to be read in relation to personal experience: immobility is an intrinsically experiential condition that can affect each individual differently and uniquely, and it is variable over time. It can only partially be explained by the objectively measurable accessibility conditions of a given spatial context. At the same time, measuring how a person moves in time and space may not exhaust the complexity of emotions, difficulties, and satisfactions derived from their immobility practices. To delve into this profoundly personal but central aspect in understanding the implications of immobility, it is necessary to consider how the framework proposed in this book can contribute to analyzing the individual's direct experience of mobility and immobility as chosen or imposed through effective analytical methodologies.

Finally, a fourth issue requiring a specific reflection concerns the contribution of the framework in providing a nuanced understanding of the potential policy implications of forms of immobility, which is useful for both territorial research and integrated urban and transport planning analysis and policies. In this context, it is relevant to consider which policy recommendations and advice can emerge from a new reflection on immobility, and what innovative policies supporting reversible immobility can arise from these reflections.

2.6 Conclusions

Discussing immobility means venturing into a complex and ambiguous terrain: in a mobile world, individuals continuously experience conditions of relative immobility. This literature review has shown how several authors from different disciplinary fields pose particular attention to the multiformity of these experiences of stillness. They focus, in particular, on the effect that barriers of various kinds, whether physical, spatial, social or cultural, can have in constraining our immobilities, often as a function of the dominant power logic which is able to mobilize some individuals and groups while immobilizing others.

As seen, it is not only the weakest individuals that are subjected to constraints producing immobilization. Similarly, the assumption that mobility differential results from subtracting a certain amount of high mobility expressed by a privileged individual to that of another less mobile and disadvantaged one is not always valid. Scientific literature suggests how the determining factor in constructing forms of immobility-related inclusion and exclusion does not depend so much on the intensity, frequency and speed of one's mobility, but on the degree of voluntariness with which an individual can choose to move or stay still without having less access to social and spatial networks than desired.

A certain level of mobility capital is fundamental in order to access resources and opportunities, and its absence (but also its 'hyper'-like expressions) can, under certain conditions, be factors of exclusion. Conversely, when this access can occur without recourse to high mobility extended in space and time, the fact of being less mobile than others but still able to obtain the desired and necessary resources represents the effect of the establishment of a new type of capital revolving around the dimensions of access and physical and spatial proximity. According to this perspective, which in scientific literature represents a junction between social and urban studies, skills are built and shaped by the spatial conditions that allow for accessibility by proximity, orienting them towards less mobility-dependent practices of appropriation. Therefore, the organization of spaces and networks assumes a relevant role in facilitating these practices and increasing *immotility* (Ferreira et al. 2017).

Understanding how various forms of immobility manifest themselves becomes conceptually pivotal in distinguishing between constrained immobilities potentially linked to social exclusion and reversible immobilities that may indicate well-being and active participation in social and spatial networks of proximity. The definitions of immobility and the framework proposed in this chapter provide insights into interpreting the varying conditions of immobility in relation to accessibility. However, they also open up new questions and issues, succinctly presented at the conclusion of the chapter that will be addressed in subsequent chapters of the book.

References

Aaditya B, Rahul T (2021) Psychological impacts of COVID-19 pandemic on the mode choice behaviour: a hybrid choice modelling approach. Transp Policy 108:47–58. https://doi.org/10.1016/j.tranpol.2021.05.003

Adey P (2006) If mobility is everything then it is nothing: towards a relational politics of (im)mobilities. Mobilities 1(1):75–94

Adey P, Hannam K, Sheller M, Pandemic DT, Hannam K, Sheller M (2021) Pandemic (im)mobilities. Mobilities 00(00):1–19. https://doi.org/10.1080/17450101.2021.1872871

Ascher F (2005) Les sens du movement: modernités et mobilités. In: Allemand S, Ascher F, Lévy J (eds) Les sens du movement. Belin, Paris

Axhausen KW (2008) Social networks, mobility biographies, and travel: survey challenges. Environ Plan B 35(6):981–996

Banister D (2008) The sustainable mobility paradigm. Transp Policy 15(2):73–80. https://doi.org/10.1016/j.tranpol.2007.10.005

Banister D (2011) The trilogy of distance, speed and time. J Transp Geogr 19(4):950–959. https://doi.org/10.1016/j.jtrangeo.2010.12.004

Bauman Z (1998) Globalization. The human consequences. Polity press-Blackwell publishers, Cambridge-Oxford

Bauman Z (2000) Liquid modernity. Polity press, Cambridge

Benenson I, Ben-Elia E, Rofe Y, Rosental A (2017) Estimation of urban transport accessibility at the spatial resolution of an individual traveler. Springer Geography 383–404. https://doi.org/10.1007/978-3-319-40902-3_21

Bertolini L (2012) Integrating mobility and urban development agendas: a manifesto. disP—Plan Rev 48(1):16–26. https://doi.org/10.1080/02513625.2012.702956

Bissell D, Fuller G (2010) Stillness unbound. In: Bissell D, Fuller G (eds) Stillness in a mobile world. Routledge, Abingdon

Bissell D, Adey P, Laurier E (2011) Introduction to the special issue on geographies of the passenger. J Transp Geogr 19(5):1007–1009. https://doi.org/10.1016/j.jtrangeo.2011.06.002

Bourdieu P (1986) The forms of capital. In: Richardson J (ed) Handbook of theory and research for the sociology of education. Greenwood, Westport, CT, pp 241–58

Bourdin A (2005) L'individualisme à l'heure de la mobilité généralisée. In: Allemand S, Ascher F, Lévy J (eds) Les sens du mouvement, Paris, Berlin

Brouwer AE, Mariotti I (2023) Remote working and new working spaces during the COVID-19 pandemic—insights from EU and abroad. In: Akhavan M, Hölzel M, Leducq D (eds) European narratives on remote working and coworking during the COVID-19 pandemic. Springer briefs in applied sciences and technology. Springer, Cham. https://doi.org/10.1007/978-3-031-26018-6_2

Cass N, Shove E, Urry J (2005) Social exclusion, mobility and access. Sociol Rev 53(3):539–555. https://doi.org/10.1111/j.1467-954X.2005.00565.x

Cervero R (1989) Jobs-housing balancing and regional mobility. J Am Plann Assoc 55:136–150. https://doi.org/10.1080/01944368908976014

Christidis P, Cawood EN, Fiorello D (2022) Challenges for urban transport policy after the COVID-19 pandemic: main findings from a survey in 20 European cities. Transp Policy 129:105–116. https://doi.org/10.1016/j.tranpol.2022.10.007

Cohen SA, Gössling S (2015) A darker side of hypermobility. Environ Plan A: Econ Space 47(8):166–1679. https://doi.org/10.1177/0308518x15597124

Conradson D (2011) The orchestration of feeling: stillness, spirituality and places of retreat. In: Bissell D, Fuller G (eds) Stillness in a mobile world. Routledge, Abingdon

Cortés-Albornoz MC, Ramírez-Guerrero S, García-Guáqueta DP, Vélez-Van-Meerbeke A, Talero-Gutiérrez C (2023) Effects of remote learning during COVID-19 lockdown on children's learning abilities and school performance: a systematic review. Int J Educ Dev 101:102835. https://doi.org/10.1016/j.ijedudev.2023.102835

Cresswell T (2010) Towards a politics of mobility. Environ Plan D: Soc Space 28(1):17–31. https://doi.org/10.1068/d11407

Cresswell T (2012) Mobilities II: still. Prog Hum Geogr 36(5):645–653

Cresswell T (2005) Justice sociale et droit à la mobilité. In: Allemand S, Ascher F, Lévy J (eds) Les sens du mouvement, Berlin, Paris

Cresswell T (2006) On the move: mobility in the modern western world, 1st edn. Routledge

de Groot JIM, Steg L (2008) Value orientations to explain beliefs related to environmental significant behavior: how to measure egoistic, altruistic, and biospheric value orientations. Environ Behav 40(3):330–354. https://doi.org/10.1177/0013916506297831

Dematteis G (2018) La metro-montagna di fronte alle sfide globali. Riflessioni a partire dal caso di Torino. Journal of Alpine Research|Revue de géographie alpine [En ligne], 106-2|2018, mis en ligne le 12 août 2018, consulté le 03 janvier 2022. http://journals.openedition.org/rga/4318; https://doi.org/10.4000/rga.4318

Dias da Silva A, Georgarakos D, Weißler M (2023) How people want to work—preferences for remote work after the pandemic. Published as part of the ECB Economic Bulletin, Issue 1/2023

Dijst M, Rietveld P, Steg L (2013) Individual needs, opportunities and travel behavior: a multidisciplinary perspective based on psychology, economics and geography. In: van Wee B, Annema JA, Banister D (eds) The transport system and transport policy. Elgar, Cheltenham

Doel MA (1999) Poststructuralist geographies: the diabolical art of spatial science. Edinburgh University Press, Edinburgh

Elliott A, Urry J (2010) Mobile lives. Routledge, New York

European Institute for Gender Equality, EIGE (2021) Gender equality and the socio-economic impact of the COVID-19 pandemic–Research note. Available at https://eige.europa.eu/publicati ons/gender-equality-and-socio-economic-impact-covid-19-pandemic

Farrington JH (2007) The new narrative of accessibility: its potential contribution to discourses in (transport) geography. J Transp Geogr 15(5):319–330. https://doi.org/10.1016/j.jtrangeo.2006. 11.007

Farrington J, Farrington C (2005) Rural accessibility, social inclusion and social justice: towards conceptualization. J Transp Geogr 13:1–12

Ferreira A, Antunes A, Pinho P (2003) Transport-sustainable urban development strategies. In: Zanon B (ed) Sustainable urban infrastructure: approaches, solutions, networking, Temi Editrice, Trento, pp 123–130

Ferreira A, Batey P (2007) Re-thinking accessibility planning: a multi-layer conceptual framework and its policy implications. Town Plann Rev 78(4):429–458. https://doi.org/10.3828/tpr.78.4.3

Ferreira A, Beukers E, Brömmelstroet M Te (2012) Accessibility is gold, mobility is not: a proposal for the improvement of Dutch transport-related cost-benefit analysis. Environ Plan B: Plann Des 39(4):683–697. https://doi.org/10.1068/b38073

Ferreira A, Bertolini L, Næss P (2017) Immotility as resilience? A key consideration for transport policy and research. Applied Mobilities 2(1):16–31. https://doi.org/10.1080/23800127.2017. 1283121

Gane N (2006) Speed up or slow down? Social theory in the information age. Inf Commun Soc 9(1):20–38. https://doi.org/10.1080/13691180500519282

Gehl J (2011) Life between buildings. Island Press, Washington D.C.

Geurs KT, van Wee B (2004) Accessibility evaluation of land-use and transport strategies: review and research directions. J Transp Geogr 12(2):127–140. https://doi.org/10.1016/j.jtrangeo.2003. 10.005

Hägerstrand T (1975) Space, time and human conditions. In: Karlqvist A, Lundvist L, Snickars F (eds) Dynamic allocation of urban space. Saxon House, Farnborough, pp 3–14

Handy SL, Boarnet MG, Ewing R, Killingsworth RE (2002) How the built environment affects physical activity: views from urban planning. Am J Prev Med 23(2 SUPPL 1):64–73. https:// doi.org/10.1016/S0749-3797(02)00475-0

Hannam K, Sheller M, Urry J (2006) Editorial: mobilities, immobilities and moorings. Mobilities 1(1):1–22. https://doi.org/10.1080/17450100500489189

Harker C (2009) Student im/mobility in Birzeit, Palestine. Mobilities 4(1):11–35. https://doi.org/ 10.1080/17450100802657947

Holden E, Banister D, Gössling S, Gilpin G, Linnerud K (2020) Grand Narratives for sustainable mobility: a conceptual review. Energy Res Soc Sci 65(July 2019):101454. https://doi.org/10. 1016/j.erss.2020.101454

ISFORT—Istituto Superiore di Formazione e Ricerca sui Trasporti (2023). 20° Rapporto sulla mobilità degli italiani AUDIMOB

Jocoy CL, Del Casino VJ (2010) Homelessness, travel behavior, and the politics of transportation mobilities in Long Beach, California. Environ Plan A 42(8):1943–1963. https://doi.org/10.1068/ a42341

Kamruzzaman M, Yigitcanlar T, Yang J, Mohamed MA (2016) Measures of transport-related social exclusion: a critical review of the literature. Sustainability (Switzerland) 8(7):6–11. https://doi.org/10.3390/su8070696

Kaufmann V (2002) Re-thinking mobility. Ashgate, Farnham

Kaufmann V, Bergman MM, Joye D (2004) Motility: mobility as capital. Int J Urban Reg Res 28(4):745–756

Kaufmann V, Viry G (2015) High mobility as social phenomenon. In: Kaufman V, Viry G (eds) High mobility in Europe. Palgrave Macmillan

Kenyon S, Lyons G, Rafferty J (2002) Transport and social exclusion: investigating the possibility of promoting inclusion through virtual mobility. J Transp Geogr 10(3):207–219

Lanza G, Pucci P (2022) Distributing, DesynchroniSing, DigitaliSing: towards a new mobile urbanity in the COVID-19 era. In: Balducci S, Armondi S, Bovo M, Galimberti B (eds) Cities learning from a pandemic: towards preparedness. Routledge

Lanzendorf M (2003) Mobility biographies. A new perspective for understanding travel behaviour. In: Paper presented at the 10th international conference on travel behaviour research, Lucerne, 10–15 August 2003

Larsen J (2014) Distance and proximity. In: Adey P, Bissell D, Hannam K, Sheller M (eds) The Routledge handbook of mobilities. Routledge, Abingdon

Le Breton E (2005) Mobilité, exclusion et marginalité. In: Allemand S, Ascher F, Lévy J (eds) Les sens du movement. Berlin, Paris

Levine J, Grengs J, Merlin LA (2019) From mobility to accessibility. Transform urban transportation and land use planning. Cornell University Press, Ithaca (NY)

Levinson DM (1998) Accessibility and the journey to work. J Transp Geogr 6(1):11–21. https://doi.org/10.1016/s0966-6923(97)00036-7

Lucas K (2012) Transport and social exclusion: where are we now? Transp Policy 20:105–113

Lucas K, van Wee B, Maat K (2016) A method to evaluate equitable accessibility: combining ethical theories and accessibility based approaches. Transportation 43:473–490

Lussault M (2005) La mobilité comme événement. In: Allemand S, Ascher F, Lévy J (eds) Les sens du movement. Belin, Paris

Lyons G, Davidson C (2016) Guidance for transport planning and policymaking in the face of an uncertain future. Transp Res Part A: Policy Pract 88:104–116. https://doi.org/10.1016/j.tra.2016.03.012

Lyons G (2014) Transport's digital age transition, March, 1–18. Mobilities 1(1):95–119

Manzini E (2021) Abitare la prossimità. Idee per la città dei 15 minuti. Egea, Milano

Marchetti C (1994) Anthropological invariants in travel behaviour. Technol Forecast Soc Chang 47(1):75–88, ISSN 0040-1625. https://doi.org/10.1016/0040-1625(94)90041-8

Marston G, Zhang J, Peterie M, Ramia G, Patulny R, Cooke E (2019) To move or not to move: mobility decision-making in the context of welfare conditionality and paid employment. Mobilities 14(5):596–611. https://doi.org/10.1080/17450101.2019.1611016

Martens K (2006) Basing transport planning on principles of social justice. Berkeley Plan J 19(1):17

Martens K (2017) Transport justice. Designing fair transportation systems. Routledge, Abingdon

Martens K, Golub A, Robinson G (2012) A justice-theoretic approach to the distribution of transportation benefits: implications for transportation planning practice in the United States. Transp Res Part A: Policy Pract 46(4):684–695

Martens K, Di Ciommo F, Papanikolaou A (2014) Incorporating equity into transport planning: utility, priority and sufficiency approaches. In: XVIII Congreso Panamericano de Ingeniería de Tránsito, Transporte y Logística, Santander, 11–13 June

Martin C (2011) Desperate passage: violent mobilities and the politics of discomfort. J Transp Geogr 19(5):1046–1052. https://doi.org/10.1016/j.jtrangeo.2011.03.005

Massey D (1993) Power-geometry and a progressive sense of place. In: Bird J, Curtis B, Putnam T et al (eds) Mapping the futures: local cultures, global change. Routledge, London, pp 75–85

Mattioli G, Lucas K, Marsden G (2017) Transport poverty and fuel poverty in the UK: from analogy to comparison. Transp Policy 59:93–105. https://doi.org/10.1016/j.tranpol.2017.07.007

Moss LAG (2006) The amenity migrants: seeking and sustaining mountains and their cultures. In: The amenity migrants: seeking and sustaining mountains and their cultures (Issue August 2006). https://doi.org/10.1659/mrd.mm008

Motte-Baumvol B, Nassi CD (2012) Immobility in Rio de Janeiro, beyond poverty. J Transp Geogr 24:67–76. https://doi.org/10.1016/j.jtrangeo.2012.06.012

Motte-Baumvol B, Bonin O, David Nassi C, Belton-Chevallier L (2015) Barriers and (im)mobility in Rio de Janeiro. Urban Studies 53(14):2956–2972. https://doi.org/10.1177/0042098015603290

Mountz A (2010) Seeking Asylum. University of Minnesota Press, Minneapolis, MN

Murphie A (2011) Shadow's force/force's. In: Bissell D, Fuller G (eds) Stillness in a mobile world. Routledge, Abingdon

Næss P (2005) Residential location affects travel behavior—but how and why? The case of Copenhagen metropolitan area. Prog Plan 63(2):167–257. https://doi.org/10.1016/j.progress.2004.07.004

Næss P (2006) Accessibility, activity participation and location of activities: exploring the links between residential location and travel behaviour. Urban Studies 43:627–652

Oviedo D, Scholl L, Innao M, Pedraza L (2019) Do rapid transit systems improe accessibility to job opportunities? The case of Lima, Perù in: Sustainability, 11

Pellegrino G (2011) Studying (im)mobility through a politics of proximity. In: Pellegrino G (ed) The politics of proximity: mobility and immobility in practice. Ashgate, Farnham, pp 1–14

Perkins A (2019) Book review. J Transp Geogr 78(2019):232–235. https://doi.org/10.1016/j.jtrangeo.2019.05.012

Preston J, Rajé F (2007) Accessibility, mobility and transport-related social exclusion. J Transp Geogr 15(3):151–160. https://doi.org/10.1016/j.jtrangeo.2006.05.002

Pucci P, Vecchio G (2019) Enabling mobilities. PoliMi Springer Brief. Springer

Pucci P, Carboni L, Lanza G (2022) Accessibilità di prossimità per una città più equa. Sperimentazione in un quartiere di Milano, in Territorio 99:40–52 ISSN 1825-8689, ISSNe 2239-6330

Putnam R (2000) Bowling alone. Simon and Schuster, New York

Pyrialakou VD, Gkritza K, Fricker JD (2016) Accessibility, mobility, and realized travel behavior: assessing transport disadvantage from a policy perspective. J Transp Geogr 51:252–269. https://doi.org/10.1016/j.jtrangeo.2016.02.001

Rau H, Scheiner J (2020) Mobility across the life course: an introduction to a dialogue In: Scheiner J, Rau H (eds) Mobility and travel behaviour across the life course. Elgar

Rau H, Vega A (2012) Spatial (im)mobility and accessibility in Ireland: implications for transport policy. Growth Chang 43(4):667–696. https://doi.org/10.1111/j.1468-2257.2012.00602.x

Rodrigue JP, Comtois C, Slack B (2016) The geography of transport systems. Routledge, New York

Rosa H (2009) Social acceleration: ethical and political consequences of a desynchronized high-speed society. In: Rosa H, Scheuerman WE (eds) High speed society. Social acceleration, power and modernity. The Pennsylvania State University Press, University Park

Santos LMD (2022) Online learning after the COVID-19 pandemic: learners' motivations. Front Educ. https://doi.org/10.3389/feduc.2022.879091

Scheiner J, Holz-Rau C (2013) Changes in travel mode use after residential relocation: a contribution to mobility biographies. Transportation 40(2):431–458

Schwanen T, Lucas K, Akyelken N, Solsona DC, Carrasco JA, Neutens T (2015) Rethinking the links between social exclusion and transport disadvantage through the lens of social capital. Transp Res Part A; Policy Plan. https://doi.org/10.1016/j.tra.2015.02.012.

Sen, A. K. (1990). Development as capability expansion. Palgrave Macmillan, London

Sen AK (1992) Inequality reexamined. Clarendon, Oxford

Shaw J, Hesse M (2010) Transport, geography and the 'new' mobilities. Trans Inst Br Geogr New Ser 35(3):305–312

Sheller M, Urry J (2006) The new mobilities paradigm. Environ Plan A 38(2):207–226. https://doi.org/10.1068/a37268

Sheller M (2018a) Mobility justice e le mobilità come bene commune. In: Perrone C, Paba G (eds) Confini, movimenti, luoghi. Politiche e progetti per città e territori. Donzelli Editore

Sheller M (2018b) Mobility justice: the politics of movement in an age of extremes. Verso Books

Silver H (2023) Working from home: before and after the pandemic. Contexts 22(1):66–70. https://doi.org/10.1177/15365042221142839

Social Exclusion Unit (2003) Report on transport and social exclusion. London

Stanley JR, Vella-Brodrick D (2011) Mobility, social exclusion and well-being: exploring the links. Transp Res Part A: Policy Pract 45(8):789–801. https://doi.org/10.1016/j.tra.2011.06.007

Straughan E, Bissell D, Gorman-Murray A (2020) The politics of stuckness: waiting lives in mobile worlds. Environ Plan C: Polit Space 38(4):636–655. https://doi.org/10.1177/2399654419900189

te Brömmelstroet M (2014) Sometimes you want people to make the right choices for the right reasons: potential perversity and jeopardy of behavioural change campaigns in the mobility domain. J Transp Geogr 39:141–144. https://doi.org/10.1016/j.jtrangeo.2014.07.001

Turner BS (2007) The enclave society: towards a sociology of immobility. Eur J Soc Theory 10(2):287–304. https://doi.org/10.1177/1368431007077807

Urry J (2000) Sociology beyond societies. Routledge, London

Urry J (2003) Global complexity. Polity, Cambridge

Urry J (2002) Mobility and proximity. Sociol 36(2):255–274

Urry J (2005) Petits mondes. In: Allemand S, Ascher F, Lévy J (eds) Les sens du movement. Berlin, Paris

van Wee B, Geurs K (2011) Discussing equity and social exclusion in accessibility evaluations. Eur J Transp Infrastruct Res 11(4):350–367

van Wee B, Geurs K, Chorus C (2013) Information, communication, travel behavior and accessibility. J Transp Land Use 6(3):1–16. https://doi.org/10.5198/jtlu.v6i3.282

Van Wee B (2021) Accessibility and mobility: positional goods? A discussion paper. J Transp Geogr 92:103033. https://doi.org/10.1016/j.jtrangeo.2021.103033

Van weeKenworthy J, Newman P (2015) The end of automobile dependence: how cities are moving beyond car-based planning. Island Press, Washington D.C.

Vannini P (2010) Mobile cultures: from the sociology of transportation to the study of mobilities. Sociol Compass 4(2):111–121. https://doi.org/10.1111/j.1751-9020.2009.00268.x

Vannini P (2014) Slowness and deceleration. In: Adey P, Bissell D, Hannam K, Sheller M (eds) The Routledge handbook of mobilities. Routledge, Abingdon

Vecchio G, Tricarico L (2019) "May the force move you": roles and actors of information sharing devices in urban mobility. Cities 88(November 2018):261–268. https://doi.org/10.1016/j.cities.2018.11.007

Vincent-Geslin S, Ravalet M (2015) Socialisation to high mobility? In: Kaufmann V, Viry G (eds) High mobility in Europe. Palgrave Macmillan

Virilio P (1999) Polar inertia. SAGE

Chapter 3
Measuring Realized Immobility

Abstract The chapter focuses on the measurement of relative immobility, building on prior research in urban studies that has employed various methodologies and tools to analyze and quantify the determinants and intensity of this phenomenon. Among the approaches reviewed—considered through the dual perspective of their limitations and opportunities—the chapter investigates the promising analytical potential of digital data, particularly mobile phone data, for measuring realized immobility. Given their growing availability and ease of use, such data can be effectively harnessed to examine individual mobility and immobility patterns. This is illustrated by an empirical study conducted in an Italian inner area, which utilizes mobile phone data to measure conditions of relative realized immobility and to explore the associations between these conditions and spatial, demographic, and socioeconomic factors. The findings indicate promising directions for further development of these research methodologies within the fields of urban and transport planning.

3.1 Introduction

The forms of relative immobility expressed in variable modes, times, and spaces by various individuals and social groups together with their potential determinants and causes, are studied by different disciplines and streams of research. The literature review presented in the previous chapter demonstrated that these aspects have also attracted interest in urban planning and transport studies, although the discussion is usually confined to a primarily conceptual dimension. This is evident in the works of authors who consider forms of relative reversible immobility as a potential outcome of the spatial and functional re-organization of settlements and transport systems in a logic of physical proximity to people and activities. At the same time, other authors take their cue from the concept of mobility as a form of capacitation, focusing on the potential for social exclusion resulting from forms of insufficient mobility linked to poor access to spatial resources and opportunities.

In the first interpretation, the promotion of reversible immobility becomes an objective that integrated land use and transport policies should endorse to increase the environmental, economic and social sustainability of cities in a proximity-oriented perspective. In the second interpretation, however, immobility can result from a lack of access to transport systems and spatially distributed activities, producing territorial inequalities and requiring contextual measures for the enhancement of potential social participation for people experiencing such forms of relative disadvantage.

Thus, the existence of conditions of relative immobility experienced by individuals living in certain territories can represent both an opportunity and a problem, necessitating the implementation of an analytical and methodological apparatus to discover and analyze the social and spatial determinants and manifestations of immobility and their implications in terms of potential social participation and inclusion. Simultaneously, such an approach can provide a valuable evaluative tool for designing and implementing flexible land use and transport policies tailored to different social and territorial circumstances. In Chap. 2, four fundamental issues have been identified regarding the translation of the concept of immobility into urban planning, the first of which pertains to the methods of measuring realized immobility. In this perspective, realized immobility is conceived as a quantifiable phenomenon that can be analyzed at both individual and territorial scales, making it suitable for an integrated approach to transport and urban planning practices. However, this is a critical point both theoretically and operationally, as the discourse on what devices and tools could be employed for this purpose is still limited. Nevertheless, some promising attempts, even if few, offer interesting starting points for a deeper reflection.

Hence, the chapter aims to provide an overview of already developed tools and methodologies to measure realized immobility and understand its determinants, addressing some of their main limitations through an empirical approach based on the use of mobile phone data in the specific territorial case study. This will be done through an initial in-depth review of previous research experiences that have actively engaged with the measurement and definition of socio-economic, demographic, and geographical determinants related to conditions of relative immobility in various case studies (Sect. 3.2). This review will identify several methodological issues that confirm how the analysis of different immobility patterns depends on effectively capturing and reconstructing complex sets of realized immobility and travel practices by deploying a plurality of techniques and tools (discussed in Sect. 3.3). Based on the knowledge gathered in the previous sections, an experimental measurement approach will be proposed in the case study of the Italian inner area of the Piacenza Apennines (Sect. 3.4) to understand the opportunities and limitations related to the use of digital data (particularly mobile phone data) for the analysis of immobility, as discussed in Sect. 3.5. Concluding remarks close the chapter.

3.2 Immobility Measurement: A Technical and Socio/ Spatial Issue

The empirical work conducted in urban studies and planning measuring immobility is part of a rather limited niche within the broader literature on mobility and transport studies. On the one hand, these works focus on immobility as a technical issue, analyzing how the survey tools traditionally used in transport planning to measure and analyze human mobility patterns, such as travel questionnaires of households, origin–destination surveys, or travel and activity reports, can record possible differentials in the respondents' travel habits and situations where travel is absent (Urbanek 2019). On the other hand, various studies have been conducted regarding the possible reasons for the absence or limited frequency and spatial extent of self-reported travel, providing explanations that focus on socio-spatial issues related to respondents' behaviors and the influence of their living and working contexts.

Regarding the aforementioned technical issues, the ability to record and measure mobility differentials, their spatial–temporal extent, and attribute them to different individuals or groups is influenced by the characteristics of the household travel and activity survey tools used to record forms of realized mobility and immobility, and by the quality of the participants' responses. Much depends, in fact, on the structure of the survey and the questions being asked. For example, many surveys on household travel and activity does not consider immobility as an occurrence that is directly surveyed and studied in its causes, but is generally inferred from the absence of self-reported travel or activities outside home (Motte-Baumvol and Bonin 2018). This last limitation applies to the Italian case (e.g., in the National Census, that mostly focuses on commuting-related displacements) while few surveys regularly conducted in other countries include variables that devote interest to immobile subjects and their daily practices (e.g., the UK National Travel Survey).

The length of the recording period is also an important factor since the shorter the temporal extent referred to for the recording of travel activities (for example, by asking a respondent to describe their displacements on a single day in the past), the greater the probability that exceptional events of immobility (illness, a day at home, vacations) can be recorded as absolute immobility, leading to an overestimation of the intensity of the phenomenon (Hubert et al. 2008). On the contrary, the longer the extent of the period, the more the results can be contextualized in a longitudinal perspective, such as to allow the identification of short-, medium-, and long-term relative immobility (Sikder and Pinjari 2012).

Another relevant aspect that may have a strong influence on the measurement of immobility is related to how travel activities are defined in the questionnaire (Motte-Baumvol et al. 2015) in terms of spatial and temporal extent. For example, Motte-Baumvol and Nassi (2012) observe, by analyzing a mobility survey in the Rio de Janeiro region implemented in 2003, that the high rates of immobility recorded in that context may depend on the fact that only walking trips made at a distance greater than 300 m from one's home were considered as actual displacements. Therefore, this survey's design implicitly excludes many proximity micromobilities, typically performed using active means, yielding inaccurate results. At the same time, not considering the extent of these movements may lead to an overestimation of the number of immobile subjects and a simultaneous underestimation of the intensity of the proximity relationships that may take place in the context of analysis. This is a significant limitation, since they represent key elements in fostering social inclusion and are possible expressions of emerging forms of proximity-based reversible immobilities.

Furthermore, the definition of the range of reasons for displacement and the possible means of transport considered by the questionnaire can also impact the quality of the analysis of forms of immobility. Limiting the survey to forms of travel linked to pre-established activities, as seen in the Italian National Census that only records work- and study-related trips, can restrict the quality and quantity of information extracted regarding individuals' multiple and complex mobility practices. Indeed, in addition to more systematic displacements, there is a wide variety of reasons that can induce other types of movements, whether related to participation in other activities, interaction with others, or, as suggested in the literature, the sheer pleasure of moving, thus challenging the idea that travel is always a derived demand (Mokhtarian and Salomon 2001). When relative immobility is analyzed and measured in the terms assumed by this research (that is, considering how, within a geographical context, individuals express mobility differentials in their daily practices) it would be important to have information on various forms of realized travel not necessarily tied to specific reasons or destinations.

Regarding the means of transport considered, excluding from the questionnaires active transport choices such as walking and cycling that are commonly used for short distances, or forms of intermodality, may, in turn, affect the quality of the analysis that measures the extent of forms of low and short-range mobilities and, by not allowing respondents to describe these movements, to identify them erroneously as absolutely immobile subjects. Both limitations could be partially overcome by proposing more detailed questions that can describe the granularity of complex mobility practices, as well as the use of new sources of information, such as those based on digital data collection and analysis or forms of digital micro-tracing. While these have several limitations, they represent a promising opportunity for measuring un-realized travel behavior (Lucas and Madre 2018; Motte-Baumvol and Bonin 2018). This aspect will be explored in more detail in the following paragraphs.

Finally, a challenging issue for those involved in survey design and those analyzing and interpreting the results is soft refusal, i.e., the non-declaration of realized trips to avoid the effort to complete the questionnaire. Madre et al. (2007) hypothesized that soft refusal has a potentially high incidence in producing results that show high rates of immobility, and Lucas and Madre (2018) pointed to the provision of incentives and rewards to induce participation and respond accurately as a possible solution to the problem.

In addition to technical issues related to the design of the tools used to measure travel activities, other scholars have focused on the determinants of immobility, explain why some spatial contexts and individual features may be more inductive to it. Both dimensions turn out to be relatively unexplored by the transport studies literature, mainly because, as seen in the previous section, more attention is given to measurement rather than to profiling, as if immobility was mainly a technical rather than a social issue (Motte-Baumvol et al. 2016; Motte-Baumvol and Bonin 2018). Again, the authors grappled with the limitations of the traditional methodologies and tools for mobility and travel estimation employed in transportation and land use planning, with the goal of profiling the less mobile subjects against other sociodemographic, economic and spatial variables inferred from mobility surveys in search of possible socio-spatial correlations that could, at least partially, explain these immobility behaviors. In these works, age, gender, employment status, availability of a private vehicle, educational level and economic condition are the sociodemographic and economic factors that explain a greater propensity to daily immobility (Madre et al. 2007; Fol 2009; Bayart et al. 2018a, b; Motte-Baumvol and Bonin 2018). Spatial features, on the other hand, are represented by population and activity density of the area of residence, the location of one's home relative to the main service-providing centers of the conurbation, and access to nearby transportation systems and activities. According to the same authors, individuals less likely to move over longer distances and more frequently are those living in areas of low settlement density with limited access to urban activities and services, the elderly or very young profiles, people who either do not work or work from home with relatively low levels of education and income, and with a more significant incidence for women. Similar results are also confirmed in the Italian case: an analysis performed within this research on a national survey about daily life activities on a valid statistical sample of respondents (AVQ ISTAT 2019) showed that mobility differentials are in place between people with different socioeconomic, demographic, and geographic-related conditions that can be related to forms of potential disadvantage. More specifically, potentially marginalized groups such as the elderlies and youngsters, women, people with low education, out of the workforce or unemployed and those with limited access to private means of transport, the internet, and perceiving generally low accessibility to basic services emerge as the populations relatively more immobile in their daily life.

From this perspective, the idea that being relatively immobile concerns people who experience potential forms of relative socio-economic disadvantage critically discussed in the previous chapter seems to be confirmed. Conversely, surveys also show that mobility levels tend to be higher in terms of frequency of travel in the center of a conurbation, but that the central location impacts the distance being traveled, which generally appears to be lower than that of those living in the suburban area, while the modal share tends to favor active mobility means or public transport. From this framework, it is possible to indicate how some forms of relatively low mobility in terms of the daily activity area's spatial extent experienced over proximity distances can be induced by good levels of accessibility, the functional mix and settlement density. In this perspective, relative immobility on higher distances, eventually coupled with high-intensity low-distance mobilities in the living surroundings, become an outcome of the voluntary organization of movements through alternative time and space strategies allowing people to remain local and should not be intended as the result of the action of specific constraint (Motte Baumvol and Nassi 2012; Milbourne and Kitchen 2014), ultimately resulting in reversible immobility.

Conversely, other forms of relative immobility may be potential manifestations of social inequality, as is the case for populations who may see their activity participation limited due to individual and spatial factors affecting their ability and capacity to access more dynamic and opportunity-rich areas, resulting in constrained immobility.

The latter hypothesis is studied in the analyses conducted on the Rio de Janeiro survey by Motte-Baumvol et al. (2012, 2016), from which it emerges how differences in daily mobility and the propensity for immobility in an exceptionally socially polarized metropolitan region are determined at the level of macro-geographic areas and depend on the structural social characteristics of the different populations. Among these, the most relevant are employment status—with higher unemployment rates in disadvantaged areas—, demographics—with a more significant presence of young and very young people in the more low-income immobile areas—and different propensities for proximity relations. Moreover, again analyzing the case of Rio de Janeiro, the authors observe that immobility can take on a geographic connotation in addition to a social dimension. Physical macro-barriers such as major road infrastructures or extensive monofunctional zones produce a severance effect that, according to survey data, determine a negative impact on the propensity for mobility of isolated neighborhoods towards the outside. However, the authors point out that for some seemingly immobile populations, confined within areas poor in opportunities and services, neighborhood life and the informal ties and jobs available represent a fundamental resource for social inclusion and survival of the individual. Participation in these informal networks constitutes social and spatial capital appropriated through micro-mobilities that analytical tools used in planning may not capture. In a completely different context, a similar condition characterizes the deprived rural areas analyzed by Milbourne and Kitchen (2014). In these low-density contexts, the lack and closures of shops, services, digital divide, and lack of public transport induce inhabitants to move less frequently but also to organize their mobility practices carefully and in more complex forms, often implementing forms of solidarity-based self-organization through travel and trip sharing and support to less mobile members of the community. These last aspects will be deepened in Chap. 5.

Hence, it is important to consider and evaluate how different analytical methods could be applied to follow and understand these complex and varied flows of immobility. Also, highly aggregated measurements of immobility, which do not consider the socio-cultural and geographic characteristics of the context, can lead to evaluation errors and biases such as the underestimation of local solidarity in contexts of apparently absolute immobility: strategies to confront social disadvantage do not necessarily require transforming potential mobility into actual mobility. Depending on the scale of analysis, locality or proximity can be mistaken for immobility (Adey 2006; Motte-Baumvol and Nassi 2012), both in apparently disadvantaged cases and in those where immobility can be reversible. This point leads us to reflect on the appropriateness of complementing a purely quantitative approach with a more articulated and disaggregated reading that can account for immobility's multiple expressions and meanings in different study contexts.

3.3 Capturing Realized Immobility Practices

The review presented in the previous section demonstrates that to identify conditions of relative immobility and their determinants, as expressed by some individuals compared to others in their daily mobility practices within a specific geographical context, it is essential to dispose of detailed information on how realized forms of mobility and immobility are expressed and differentiated.

Traditional methods of analyzing travel behavior and flows used in transport planning have generally relied on the manual collection of quantitative data that are useful for estimating travel demand, service level and transportation supply to build predictive models for policy-making purposes and transportation planning and management (van Wee 2013; Milne and Watling 2019). The primary tools employed in this regard include household questionnaire surveys, origin–destination surveys and travel diaries. In addition, direct registration methods are also employed, such as surveys of passengers and traffic flows, usually implemented through sensors (Urbanek 2019). These tools, thanks to the variety of questions asked in the survey questionnaires, allow for the collection of information on the sociodemographic characteristics of respondents, which is useful for profiling their travel behaviors. However, the complexity and the cost of implementing these surveys make them somewhat rare and discontinuous, providing a static picture of travel habits at the aggregate level of a population along with a generally short registration period and based on questions covering a limited range of reasons and means of travel (Thakuriah et al. 2017). In addition, questions have been raised regarding the reliability and representativeness of data resulting from the selection of sample respondents that allow for spatially and demographically aggregated surveys. Other debated aspects concern the actual strategies of contacting and recruiting participants, co-inviting them to obtain reliable responses, and the design of the questionnaires, which the respondent should easily fill without compromising the instrument's validity. Finally, other critical points regard the need to make analysis and modeling tools sufficiently flexible to work with different data and to be capable of accounting for the complexity of modes and times of travel (Stopher and Greaves 2007; Bayart et al. 2018a; Richard and Rabaud 2018).

As for the last point, it is relevant to note how a focus on demand for mobility expressed as a flow of displacements tends to encounter significant limitations when dealing with contemporary urban mobilities, their spatial reflections and the social dynamics they generate risking to overlook more profound social implications of movement (Pucci and Vecchio 2019b, p. 6). Such a simplified consideration of mobility as a pure physical movement linking an origin to a destination, explained by a specific reason, and supported by one or more means of transport according to rational logic depends on the need to develop analytical methodologies that constitute a toolbox of well-established, ready to-implement, and easily transmissible approaches, thus operationally usable in the field of planning (Pucci and Vecchio 2019b, p. 1). At the same time, it seems increasingly important to update this toolbox by identifying new modes of analysis delving into the complex and often ambiguous social, cultural, and political meanings assumed by our ways of experiencing the physical and imaginative space in which we move. Such an awareness asks for new research paths that unpack the interaction between transport and society to help figure out ways of moving less, or at least differently, and that makes the disciplinary field of transport geography and planning to be a more human geography (Shaw and Sidaway 2010; Schwanen 2016). Considering mobility as a complex set of social and spatial practices, a perspective that was first advocated by the proponents of the 'mobilities turn' in the research stream of mobilities studies, means employing a wealth of empirical multidisciplinary research techniques that can innovate the social sciences and address the intertwined practices of many different kinds of contemporary (im)mobilities at a variety of speeds and scales (Sheller 2014, p. 12). It has been proposed that these creative techniques, defined as mobile methods, must not only be able to capture but also keep pace, simulate, mimic, reconstruct through objects, film footage, collection of direct experiences, and 'go along with' through participant, performative, ethnographic techniques the movements, blocked movements, potential movement and immobility, dwelling and place-making (D'Andrea 2006; Büscher and Urry 2009; Büscher et al. 2010; Merriman 2014; Bissell 2018; Vannini and Scott 2020). As promising as they are in exploring the more human qualities of mobility (Shaw and Hesse 2010, p. 309), mobile methods are primarily conducted through research techniques that cannot rely on numerically significant populations and are often not structured for robust analytical protocols that could be verified, communicated, and replicated (Pucci and Vecchio 2019b, p. 6). The main consequence is that, compared to the approach and tools of transport geography and transport planning needs, mobile methods are less policy-driven because they are perceived as lacking a solid, empirically-based assessment of how mobility works (Shaw and Hesse 2010, p. 308).

However, it would be superficial to view these different approaches as impermeable and rigidly defined. This mindset overlooks the diversity of research being undertaken in both mobilities studies and transport geography and planning, and that often demonstrates a mutual influence in the mixed and innovative methodologies that have been proposed in recent years in the name of a beneficially mutual dialogue over more topic- or problem-oriented debates on the (non)movement of people and objects (Bissell et al. 2011; Shaw and Sidway 2010; Merriman 2014; Schwanen 2016). In this perspective, several authors advance the idea that a promising point of contact between the need to provide robust and representative aggregated mobility analysis to be included within research and policymaking protocols and the recognition that any methodological simplification risks resulting in the neglect of the complexity of mobility, is provided by the increasing availability of the so-called big digital data (DeLyser and Sui 2012; Schwanen 2017; Pucci and Vecchio 2019b). Digital data obtained from different sources such as mobile phones, GPS, public transport smart cards and social networks offer a precious opportunity to understand the characteristics and evolution of space–time mobility and immobility practices of people and the patterns of relationships they support at a fine grain, shedding light on complex social phenomena and becoming indicators of the quality of transport systems, the functional diversity of urban contexts, and the quality of public spaces (Ahas et al. 2010a, b; Kitchin 2014; Rabari and Storper 2015; Schwanen 2015; DeGennaro et al. 2016; Kourtit et al. 2016; Docherty et al. 2018; Milne and Watling 2019; Welch and Widita 2019). The availability of information collected at the individual scale, both passively and by the direct involvement of people through crowdsourced and volunteered data collection (Goodchild 2007; Brabham 2009), makes it virtually possible to investigate at the same time microscopic personal choices and macroscopic behaviors (Chen et al. 2016). A significant consequence is the possibility to combine different scales of analysis to bring out different immobility profiles and relate them to the multiplicity of preferences, needs, contextual features and personal abilities that influence them.

It is also argued that passive and active digital data collection systems can, when fully operational, make the processes of collection, analysis and forecasting adopted in transport planning, traditionally based on a stated preference approaches and manual data collection, more efficient and cheap, allowing planning to become more contingent and to promote, in perspective, better and potentially continuous management of transport systems and other urban services (Concilio et al. 2019; Urbanek 2019). At the same time, digital data could assume a relevant role in urban governance, contributing to the response to emerging issues of various kinds, dealing with variable timeframes and with different possible approaches. According to Semanjski et al. (2016), data can support strategic activities related to long-term mobility planning measures; tactical decisions, providing information for the implementation of strategic activities; operational decisions in the form of rapid responses to everyday problems in real-time perspective.

Scientific literature, however, focuses on the many critical aspects related to data use in the fields of mobility research and policy. This is the case with privacy issues, which is particularly relevant for mobility analysis, where tracking individual movements becomes a valuable analytical resource (Boyd and Crowford 2012; Kitchin 2013). Another aspect to be considered concerns the complexity of the use of data in public policies and the necessary technical/political expertise that is a prerequisite for extracting value from them. In mobility-related decision-making, this aspect translates into the readiness to collect, manage and analyze different types of data that can support the development of policymaking processes. Since mobility practices are extremely complex and must also be considered in relation to the context in which the policy is proposed (Aguiléra et al. 2012), it may be necessary to consider the actual representativeness of the collected data and eventually complement it using data of multiple types of digital and non-digital origin obtained from different sources, selected according to the specific mobility-related phenomenon that is to be analyzed. This issue requires significant technical, economic and managerial capacities to handle and integrate different datasets owned by different providers (Milne and Watling 2019; Tranos and Mack 2019; Lanza 2021) to assess their completeness, quality and usefulness to the problem to be addressed and to involve stakeholders (from data owners to recipients of a specific policy measure) who can play a key role in the success of the process. These and other difficulties are related to the fact that ICTs and data collection/analysis methods are still being developed and are a probable cause of the low interest expressed by transport planning in the opportunities offered by digital data (Vecchio and Tricarico 2019).

However, although the limitations described above introduce unprecedented challenges from both a theoretical and a practical point of view, it is also true that digital data can complement and update the research tools traditionally used in the fields of transportation studies and mobilities, enriching our way of seeing and understanding the complexity of immobility practices. For example, Kitchin (2014b) argues that the future development of a 'data-driven science' in which digital data will support research and analysis of known problems is likely. However, this does not mean that big data will subvert completely the way we analyze complex urban dynamics. Indeed, not all phenomena can be measured through digital data analysis (McNeely et al. 2014), raw data will still need to be processed through algorithms, calling into question their objectivity, and the cause-effect correlations that can be identified through data still need to be verified (Rabari and Storper 2015; Kwan 2016). Established small-data based techniques and methodologies will still be relevant and should be coupled with the new big-data driven approaches that are being developed. In particular, it is essential to understand how to combine and interoperate big digital data with both small quantitative and qualitative data. While standards of interoperability between traditional quantitative and digital data already exist, the combination of qualitative and digital quantitative methodologies remains of great interest because the former provides a window into the feelings of individuals, their experience and their interpretation of practices of which digital data are the trace and which are often difficult to understand otherwise (Lucas 2020). Such awareness is particularly relevant for analyzing both forms of mobility and immobility.

3.4 An Experimental Approach to Immobility Measurement Through Mobile Phone Data Analysis

This section presents an experimental approach to immobility measurement using mobile phone data conducted in the mountainous areas of the province of Piacenza (Italy). The territorial case, encompassing an area of approximately 2200 km^2 and covering the central and southern portions of the province of Piacenza, is characterized by particular orographic conditions. While the northern part of the province is predominantly flat and hilly, featuring several main centers and a networked road system, the southern portion is typically mountainous and develops along two main valleys (Val Trebbia and Val Nure). In this sparsely populated area, the main centers are mostly located in the valley bottoms along the main roads, surrounded by numerous small, dispersed hamlets at higher altitudes. In recent decades, like many other inland areas of the country, significant shrinking processes have affected the mountain villages in the southern part of the province. These villages have experienced a depopulation process mainly induced by internal migration towards more dynamic contexts within the province or the country, only partially countered by the arrival of young inhabitants, the return of older ones after retirement, and the strenuous resistance of those who have always remained here (SNAI report 2018).

As a consequence, the municipalities of the southern valleys today are character-ized by an aging population, low income and educational levels, unemployment, and work/study-related mobility practices over long distances, revealing low levels of attractiveness and significant dependence on more dynamic areas of the region (Vendemmia et al. 2021). These trends also explain why some of the municipalities in the high valleys were classified as peripheral according to the Italian National Strategy for Inner Areas (SNAI), with low levels of accessibility to a set of basic services including railway transport, higher education, and healthcare.

The study case of the Piacenza Apennine was selected because it exemplifies territorial conditions characterized by specific fragilities recurrent in the national context. Many of the profiles overrepresented here, particularly people out of the active population, elders, poorly educated people, and low-income residents, are generally associated with lower rates of daily mobility, as discussed in the previous section.

In this perspective, analyzing immobility conditions allows shedding light on the dynamics affecting this territory, verifying the specific emerging conditions, and identifying whether the trends statistically identified in the literature are confirmed here.

To identify and measure the daily mobility differentials of the inhabitants in the Apennine area of the province of Piacenza and address the limited analytical opportu-nities offered by the traditional census data available in Italy, the research utilizes data derived from raw mobile phone records purchased from the main Italian telecommu-nications company (TIM). Among the vast array of digital data that can be fruitfully employed in urban and territorial research discussed in the previous section, mobile phone data were selected due to their ability to record movements and static pres-ence patterns in time and space of a large sample of individuals, collected over wide study areas and considering very extended timeframes (Blondel et al. 2015; Tu et al. 2017; Manfredini et al. 2022). When connected to antennas that support the cellular network, mobile phones produce MPRs (mobile phone records), returned as times-tamped data on users' location which telephone companies actively or passively collect for billing purposes (Ahas et al. 2010a; Urbanek 2019). MPRs are structured as chains of movements in which the positioning of a single user is referred to the nearest telephone antenna and measured at very short intervals (Ahas et al. 2010b), ensuring a tracking capability that records the activities of potentially representa-tive samples of users for research due to the high penetration rate of mobile phones among the population.

Due to their flexibility and completeness, mobile phone data enable four types of analysis and sensing approaches indicated by Pucci et al. (2015, 2022). The first one, defined as *mobile landscape method*, uses tracking data to understand patterns of everyday life in the city and assess the distribution of urban activities in different temporalities (Calabrese et al. 2013; Jiang et al. 2016). The second approach, called *social positioning method*, profiles users by relating movement patterns to their social and spatial interactions (Ahas et al. 2010a, b; Picornell et al. 2015). The third one, defined as *urban spaces classification*, considers mobility behaviors in relation to land use characteristics by assuming a user's patterns of movement as good indicators of the city's functional composition (zoning) and the activities present (Tu et al. 2017; Ni et al. 2018). Finally, a fourth approach exploits the information brought by mobile phone data tracking to estimate traffic flows and develop analysis and monitoring tools for managing transportation and road systems (Chen et al. 2016).

As can be seen, the first three approaches outlined by Pucci et al. (2015, 2022) exploit some of the potentials of mobile phone data for the analysis of the variation of travel flows and static presence over time expressed by different user profiles, even relating users' behaviors to the settlement, morphological and functional characteristics of a territory. On a theoretical level, thanks to their characteristics, mobile phone data can offer a more fine-grained and complete reading of phenomena that are very difficult to explore through the analysis of traditional quantitative data such as population censuses or household travel and activity surveys. The opportunity provided by the Department of Architecture and Urban Studies at Politecnico di Milano to get access to and analyze a rich dataset of MPR data tracking spatial–temporal human presences in the case study area with options for user profiling was seized founded on the expectation that these resources, which are difficult to access due to the costs and technical skills required, can contribute significantly to the advancement of this research. To this end, an analytical approach was outlined and progressively adapted to address the limitations and difficulties that emerged during the process and the inherent rigidities of data collected for purposes not directly related to the study of forms of immobility at the territorial level.

The MPR data made available by TIM record the number and variation of human presences within a geographical area that includes 29 municipalities in the Apennines and foothills areas of the province of Piacenza (Fig. 3.1). Due to the data structure, the municipal level is the minimum (and smaller) territorial unit of geographical analysis available for data collection.

Within the 29 municipalities, chosen based on their socioeconomic and demographic characteristics, approximately 100,000 people live according to 2018 ISTAT data. Based on the contract signed with the telephone company, this is the maximum number of people that can reside within a territorial 'scenario' of data collection and analysis. This limit also explains why additional municipalities in the plain area or outside the provincial boundaries were not included, preferring to focus on the dynamics affecting the Apennine villages and the closest urban centers of the foothills area that have direct relations with the former to allow for comparison between different coexisting territorial conditions.

Data is collected by the company and made available over an extended period that can reach up to 15 consecutive months. The human presence values are counted through MPRs produced by TIM customers' mobile devices within the municipal areas at 15 min intervals and returned as the hourly average of the four measurements. To extend the measurement sample from customers to the total population, TIM uses a multiplication coefficient sized according to the company's market share in the various territorial contexts to obtain values that are as close as possible to the actual situation. In this way, mobile phone data allow analyzing the variations in human presence within each geographical area throughout the selected day, month and year. In addition, records of presences are subsequently processed through algorithms by

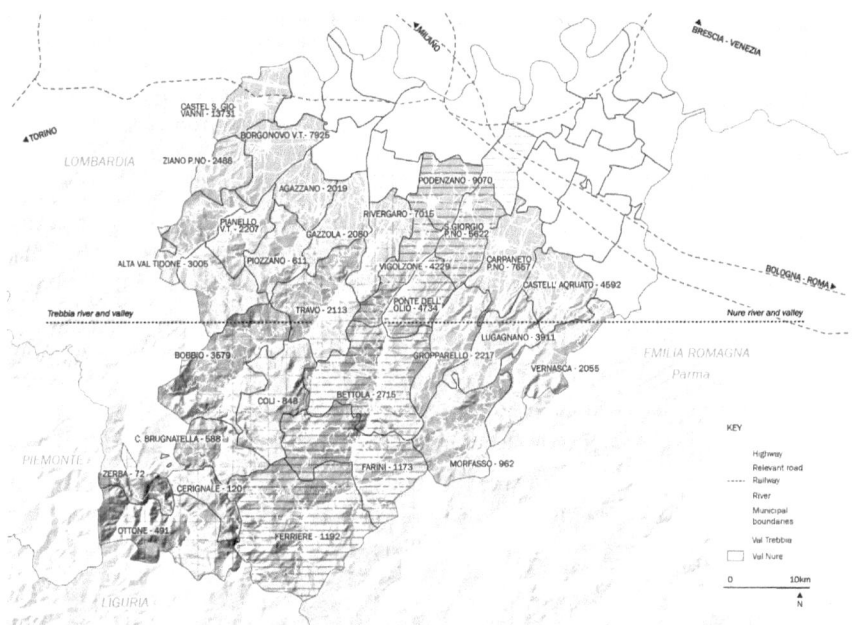

Fig. 3.1 Municipalities included in the TIM mobile phone data collection

the telephone company to provide specific user profiling options validated on the total population using coefficients based on ISTAT population data. These concern both demographic profiling (gender, age group deduced by the contracting information) and other information related to the type of connection of the user with the area. In this last case, the profiling allows defining if a user is a resident of the area based on their constant presence in the municipality during nighttime (06–22 h) in the previous thirty days. On the contrary, a user will be classified as a commuter if usually detected in the municipality only during daytime hours on working days. TIM defines users unrelated to the municipality where they are registered as visitors according to their place of origin (same region, outside the region, abroad). In general, the profiling options allowed by this type of data overcome some intrinsic limits of the statistical data available for the context. Indeed, they allow recording the actual presence of people, not only considering the registered inhabitants, and understanding the hourly and seasonal variations in the intensity of territory use in its different parts (Lanza et al. 2022; Manfredini et al. 2022).

The empirical approach work required a 15-month data collection scenario that included a pre-pandemic 'normal' phase, a total lockdown phase in Italy from March to May 2020, and a subsequent period of gradual reopening. Thus, data were collected from July 2019 to September 2020 for the entire survey area with hourly detail. However, in using such data for this research, specific criteria were adopted that led to the final selection of a more circumscribed period. First, the selected time interval should be sufficiently comprehensive to show stable trends possibly not affected by momentary and unusual fluctuations but, given the amount of data available, manageable in terms of computational capacity. Second, it was decided to consider a period before the outbreak of the Covid-19 pandemic to describe an ordinary situation without the effect of forced immobilization that the pandemic created for many months. Although the impact on the forms of daily immobility induced by the pandemic indeed represents an interesting topic to delve in, it lies outside the scope of this research mainly because of the absence of usable data referring to the waves that occurred after September 2020, which could have further erratically modified the daily rhythms of vast areas of the Italian territory. Third, it was chosen to consider a working period outside summertime. This choice was made to limit the effect of recording short- or medium-term tourist presences, concentrated in the summer months and weekends (Lanza et al. 2022) and focus on permanent residents who use local services and opportunities throughout the year. Fourth, the choice was made after a careful analysis of the completeness and reliability of the MPR for the five months that meet the above criteria. As will be seen below, the measurement sometimes returned incorrect results or, due to technical problems, very limited or absent recordings, placing further limitations on the choice of the period, which finally concerns the month of November 2019.

Hourly Data for the period 01- 30 November 2019 were directly obtained from a DVI visualization platform created by TIM at the request of the technicians of the Maudlab of the Department of Architecture and Urban Studies at Politecnico di Milano. The platform's features resulted not sufficiently flexible and complete for the type of objective assumed by the research. To overcome this limitation, the raw data files for the 15 months were downloaded directly from the platform allowing for all the filtering options available to be analyzed through data analysis software. In addition, at this stage, a filter that is featured in the DVI platform was applied to download only the values of presence related to residents and associated demographic profiling data (gender and age group), thus purposely not including in the subsequent analysis commuters or visitors. This delimitation of the sample is functional to the analysis of immobility of the inhabitants of a territory since it excludes from the calculation other people present at that specific time but who do not have, according to the criteria and analysis defined by TIM, a continuous relationship of residential type. Subsequently, the datasets were subjected to careful control and application of the criteria for selecting the most appropriate period for the analysis. The control process served primarily to map the completeness of the information at the level of data availability in the different municipalities (geographic validity) and the duration of the tracking period (temporal validity) and to identify the possible presence of outlier values. The control phase also included the comparison between the average number of human presences of residents registered according to the TIM methodology in each municipality during the whole period between 10 pm and 6 am and the value of the official record of the resident population according to ISTAT (2019 data). The comparison shows a general consistency between the two resulting figures in the orders of magnitude, although these can be compared primarily by considering the trends depicted by the curves rather than the actual values recorded. Nevertheless, mobile phone data record usually returns higher values than statistical information.

The upward trend affects almost all foothill municipalities, while opposite situations are recorded in some Apennine centers where the population value of the statistical data exceeds that recorded by TIM. An explanation for these patterns may lie in the forms of frequentation of the territory: if, on the one hand, the economically dynamic municipalities of the plain and foothills areas can attract new temporary inhabitants for working reasons that may not be officially registered but detected by telephone antennas, so some officially registered inhabitants in the Apennine municipalities may not spend all the year in these areas preferring other domiciles. At the end of the process, 541,551 hourly resident presences measurements were thus isolated during November 2019 for all 29 municipalities and including all possible profiling combinations.

Based on the characteristics and potential of the TIM data, it was decided to measure the propensity for immobility of the inhabitants (used here as a synonym for resident, following the definition provided by TIM) living in each territorial unit of analysis (municipality) by relating it to the magnitude of changes in the number of residents recorded during the period considered. This approach assumes that a decrease in the number of residents registered between two successive measurements at the municipal level indicates an outgoing flow, from the boundaries of the municipality, of a certain number of inhabitants directed towards other areas. Conversely, an increase may indicate an incoming flow to the municipality by those outside it. In both cases, the recorded change represents a proxy for the rate of mobility and displacement to and from the municipality by local inhabitants: the more the number of presences of residents changes significantly, the more likely it is that inhabitants are engaged in mobility practices requiring physical displacement to other municipalities and areas of the territory. Conversely, places inhabited by less mobile populations will be those in which the rate of change recorded by the data is lower, indicating a greater propensity for self-containment within the municipality. The entity of the variations in human presences was calculated as the standard deviation of the detected hourly presence of residents for each unit of geographical analysis, an operation that allowed measuring the variability of the values inside the sample and how they deviate from the monthly average (Salkind 2007, p. 973) assumed as the baseline using the following equation:

$$\sigma = \sum \frac{(x - \bar{x})^2}{n - 1}$$

where σ is the standard deviation, x is each individual value, $\bar{x}$ is the mean of all the registered value, n is the sample size.

It follows that the smaller the standard deviation, the smaller the average distance each data point has from the mean of the distribution and, therefore, the variation of presences, indicating conditions of relative immobility in terms of extra-municipal movements compared to other municipalities in the study area.

The calculation of the standard deviation, however, does not return a result that is comparable among different units of geographical analysis because of the various orders of magnitude of the values of human presences that characterize municipalities with highly variable population data (from 13,731 inhabitants of Castel San Giovanni to 73 of Zerba, ISTAT 2018). The application of a coefficient of variation, a measure of the spread of a dataset defined as:

$$cv = \frac{\sigma}{\mu} 100$$

where σ is the standard deviation and μ is the mean of the sample, allows normalizing the data to overcome this limitation and return a result that facilitates comparability between different areas of the territorial context of analysis.

The analysis of the presence variations resulted in the im-mobility index, which was initially calculated regardless of gender or age group. Subsequently, the level of detail was increased by filtering the data based on demographic characteristics to evaluate the differences between the various profiles of inhabitants.

In a first step, the im-mobility index was calculated for each municipality, normalized into five ranges through the Jenks optimization system defining five possible conditions (very low, low, medium, high, very high immobility), and mapped at the territorial level to identify its spatial distribution and show any differences between contexts that are heterogeneous from the settlement and socioeconomic point of view. The results show how the most significant values of individual daily mobility characterize the southern mountain area and more peripheral municipalities (Fig. 3.2). Lower values, but still significantly higher than the territorial average, are also recorded in some intermediate municipalities between the mountainous and hilly areas, as well as in some of the villages located close to Piacenza. On the other hand, medium–low or very low mobility values are recorded in the band of urban centers located in the northern and more infrastructure and economically dynamic plain area. The only municipality that differs substantially from the neighboring centers of the upper mountain valleys is Bobbio, showing a minimal average variation in the number of residents, making it the least-mobile municipality in the Apennine area.

Further interesting is the distribution of the variations in presence considering the gender of the users tracked by TIM. Figure 3.3 shows the variation of female residents compared to males in the registering period. In most municipalities, women are the most mobile population, with particularly significant peaks in the southwestern mountain villages. The only four municipalities where male residents seem to be more active in displacing (marked by negative values) are unevenly distributed between the Apennine area (Bobbio) and the foothills.

In the left column of Fig. 3.4, the levels of variation for each age group are mapped at the municipal level while, in the right column, maps depict the ratio between the intensity of variation measured for female and male individuals to show which gender profiles are more or less mobile within each age group. This analysis shows that, over the recording period, the inhabitants of the most peripheral municipalities among those in the Apennine range were generally more mobile daily than their counterparts living in the hilly and plain areas of the province. This trend is particularly evident for the residents under 18 y.o., who represent one of the most mobile segments of the population throughout the analysis area and within each municipality, and over

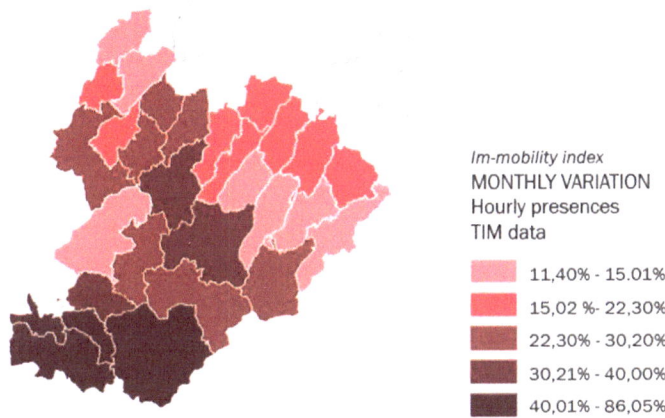

Fig. 3.2   Im-mobility index: presence variation during November 2019, total

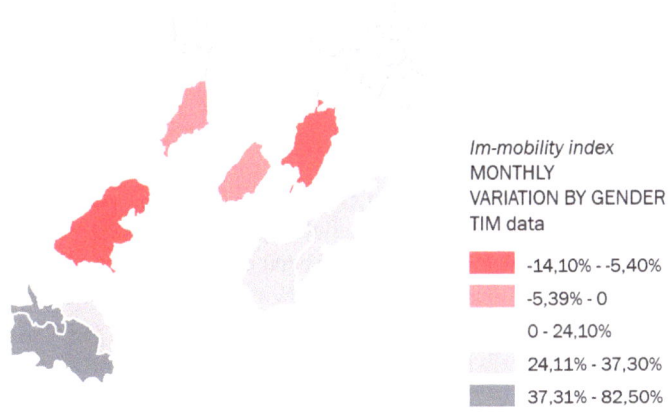

Fig. 3.3   Im-mobility index: presence variation during November 2019 by gender where negative values indicate a higher deviation in male residents' presences compared to females and viceversa. *Note*: In Fig. 3.3, red tones denotes areas where the variation of male residents is higher than female. Grey tones denote the opposite trend

30 y.o. At the same time, considering the 18–30 range, the rate of variation recorded is relatively medium–low and does not differ significantly among the municipalities of the Apennine area. The behavior of residents in the 30–40, 40–50, and 50–60 age groups (working-age bracket) tends to be relatively similar, with the inhabitants of the municipalities of the high valleys and some intermediate centers frequently moving from their municipality. More pronounced, however, is the distribution of variations concerning the over 60 segment, which is generally mobile or very mobile

in most mountain and most peripheral municipalities. On the other hand, Bobbio and the plain city of Castel San Giovanni are among the municipalities where the number of inhabitants within the territory tends to remain generally stable; in the case of Bobbio, it is also the one where men are more mobile, against a widespread and constant trend within the area that in most municipalities sees women as the most mobile component of the population for almost all age groups.

3.5   Contributions and Limits of Mobile Phone Data for Immobility Practices Recording

The analysis of daily immobility using mobile phone data has revealed territorial phenomena that could not be deduced from traditional census data and are, at least partially, unexpected. In particular, the im-mobility index shows how relative immobility is not a condition that characterizes the Apennine and peripheral municipalities of the area but that, on the contrary, affects places that emerge as more dynamic and less disadvantaged from the socioeconomic point of view and in terms of the availability of services and opportunities. At the same time, and again in contrast to what emerged from the literature review, women turned out to be the most mobile component of the population. Furthermore, no substantial differences were observed in terms of propensity for immobility related to the age of the inhabitants.

From this perspective, the substantial variations of presences homogeneously distributed among all demographic and gender profiles challenge the notion that the most peripheral and fragile areas of a territory can also be the most immobile ones, not only from a social but also from a spatial point of view. This result confirms the idea that the link between immobility and social exclusion can be at least partially reconsidered, leading to the emergence of new possible forms of mobility-related (and not immobility-related) disadvantage. Such consideration is particularly true for those areas of the territory where being relatively highly mobile for a resident becomes a basic necessity for subsistence in the presence of low levels of accessibility to local services and opportunities.

The research approach based on mobile phone analysis has also shed light on several aspects related to the potential contribution of these sources of information for scientific research on marginal low-density areas, their fragilities and the existing spatial–temporal dynamics of human presence and immobility. More in detail, it emerged how digital data could support the analysis of territorial phenomena that are difficult to identify and investigate by only relying on traditional information sources based on 'small' data, providing new valuable insights.

However, the approach suffers some intrinsic limitations of mobile phone data, which may compromise their usability in research exploring complex immobility patterns.

The first relevant limitation concerns the relationship between a research question and the possibility that the data, as they are collected and made available by owners

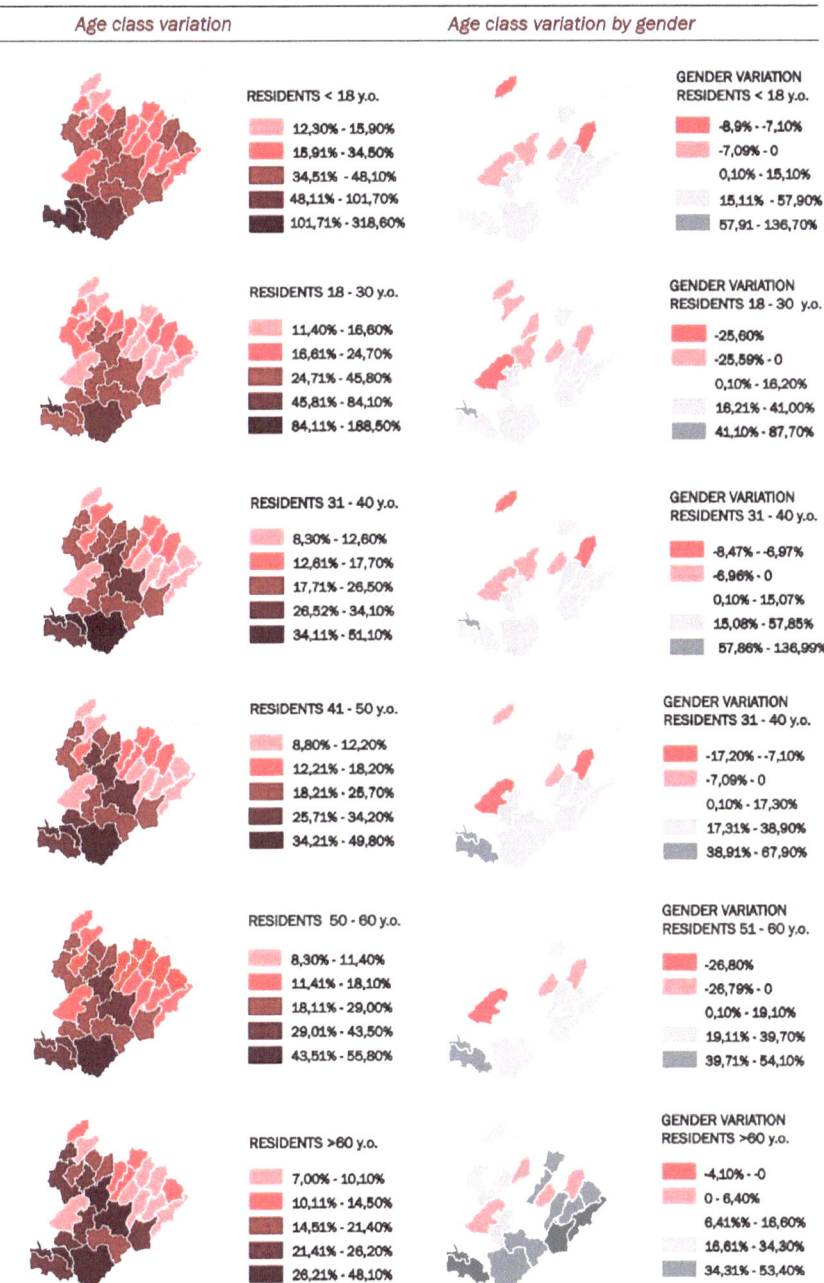

Fig. 3.4 Variation based on age class (left column) and related gender ratio (female/male) where positive values means that variation in female populations is higher than male's

and providers, can offer valuable and rigorous answers. The problem has already been raised by Miller (2010) and is related to the fact that digital data (expensive to generate, process and store) are produced in very controlled ways using sampling techniques that limit their scope, size and temporal detail. For this reason, the possibilities for researchers to customize and query preconfigured data provided with some easy-to-use basic features (as the DVI platform created by TIM) are minimal, making them not always adequate to answer complex research questions, except through intensive processing or data fusion (Calabrese et al. 2010). Hence, exploring the phenomena of interest in the case of the Piacenza Apennines required first to construct a methodological framework that would allow extracting from the data as much information as possible could be beneficial to the objectives of this work in the awareness that it was not explicitly collected and provided to answer the initial research questions. This explains why, the obtained results return the patterns of immobility only in a limited and approximate way, offering a necessarily general and aggregate picture, both from the point of view of the social sample and of the spatial–temporal detail. Therefore, this exploration helps identify some general dynamics at the territorial level but is much less significant for analyzing the variety and characteristics of the multiple behaviors of immobility that can be expressed at a scale closer to the individual, whose in-depth study clearly requires other research methodologies and analyses. Moreover, it is also important to note that these data cannot define the intensity of the recorded mobilities from a spatial and temporal point of view, since they assume static geographical units of analysis rather than focusing on the nature of the recorded displacements.

A second limitation relates to the low spatial flexibility of the data and the high level of aggregation available. As telecommunications operators collect mobile phone data unrelated to scientific research, the data being collected and processed are not entirely suitable for spatially disaggregated analyses at the submunicipal scale. At the same time, these data offer a somewhat limited spectrum of options for sociodemographic profiling (Calabrese et al. 2010). Whereas, in the latter case, in-depth analysis of individual characteristics and behaviors would raise critical issues due to privacy concerns and the limited scope of socio-demographic data collected by practitioners, increasing the spatial disaggregation of data represents a more concrete opportunity. In the Piacenza Apennines case study, an attempt was made with TIM to measure human presence at the sub-municipal scale to explore its temporal variations in both rural areas and sparsely located hamlets. If this attempt had been successful, it would have represented a very promising way to overcome the effect given by the rigidity and the high level of aggregation due to the use of municipal administrative boundaries to delimit the data collection areas. However, given the technical complexity of the request, the attempt has not returned valid results, limiting the possibility of identifying forms of micromobility of proximity that can take place within municipal boundaries and, in turn, represent important opportunities for activity and social participation.

Finally, a third limitation concerns the reliability of the mobile phone data collected and processed in a low density and aging context as that of the Apennines of Piacenza. Compared to dense urban environments, rural areas present a sparser distribution of antennas (the basic infrastructure to infer user location), with the risk of underestimating the movements of individuals and returning spatially incorrect results (Williams et al. 2015) in the municipal-level redistribution of hourly tracking. Such a problem is reflected in measurement gaps and outlier values, especially in more remote and less populated municipalities, which the algorithm used to translate the raw data into geospatial position is not always able to correct.

Also entrusted to recalculation algorithms is the quantification of human presences based on the market share of the telephone company, a step that can, in turn, create bias linked to the uneven territorial distribution of customers (Salat et al. 2019). This aspect is also reflected in the representativeness of different age groups, especially in territories inhabited by high percentages of elders. However, this latter bias seems to become less relevant thanks to the steady growth of mobile phone penetration in all age groups in Italy and other countries worldwide (Deloitte 2019).

The encountered limitations, even if significant, can be overcome or, at least, mitigated by fruitful collaboration and negotiation between the data provider and the research institution. Undoubtedly, mobile phone data still represent a relevant opportunity to be further explored to analyse 'hidden' spatial–temporal practices not only in dense urban settings, but also in dispersed, low-density territorial contexts. Nonetheless, the experience of this research suggests that the data used can only provide some additional elements to the surveys traditionally employed in mobility analysis, albeit relevant, and that the results of these analyses must be combined with information sources of a more traditional nature, necessary to contextualize and interpret the spatiotemporal patterns emerging from the data (Alexander et al. 2015).

3.6 Conclusions

Capturing immobility practices and exploring their social and spatial determinants is a challenging endeavour, as demonstrated by the limitations of many measurement experiences and tools cited in this chapter. The primary challenge lies in developing suitable methodologies and tools to effectively illuminate these practices, their spatial and temporal evolutions, and the individuals and groups who engage in them. The experimental approach proposed here relies on a methodology that exploits the opportunities of mobile phone digital data as an analytical technique capable of overcoming some limitations inherent in more widespread measurement experiences and producing results that offer a multiscalar perspective. This perspective ranges from an aggregated territorial dimension suitable for urban and transport planning activities to aspects closer to the individual sphere, thanks to profiling options and a high level of socio-spatial detail. In the case study of the Piacenza Apennines, this approach led to unexpected results regarding the social profiles and geographic areas that emerged as more or less mobile than traditionally assumed in the literature.

The discussion of the opportunities and limitations resulting from this empirical approach has identified valuable insights on how to improve mobility and immobility measurement tools and methodologies while observing and critically rethinking the implications of immobility and its possible association with forms of exclusion and socio-spatial marginality. At the same time, it unveiled the limitations of a quantitative methodology that assumes immobility as the mere absence of travel flows, unrelated to the wills, choices, and spatial constraints that affect individuals and shape their behaviors. The analytical and policy-related utility of these approaches lies in constructing specific territorial frameworks to guide further investigation into specific situations that emerge from the same analyses. Therefore, it is suggested that aggregate approaches be complemented with other analytical techniques that consider spatial conditions enabling reversible immobility and the variety of multiple immobility behaviors expressed at a scale closer to the individual.

References

Adey P (2006) If mobility is everything then it is nothing: towards a relational politics of (im)mobilities. Mobilities 1(1):75–94. https://doi.org/10.1080/17450100500489080

Aguiléra A, Guillot C, Rallet A (2012) Mobile ICTs and physical mobility: review and research agenda. Transp Res Part A: Policy Pract 46(4):664–672. https://doi.org/10.1016/j.tra.2012.01.005

Ahas R, Silm S, Järv O, Saluveer E, Tiru M (2010b) Using mobile positioning data to model locations meaningful to users of mobile phones. J Urban Technol 17(1):3–27

Ahas R, Aasa A, Silm S, Tiru M (2010a) Daily rhythms of suburban commuters' movements in the Tallinn metropolitan area: case study with mobile positioning data. Transp Res Part C 18:45–54

Alexander L, Jiang S, Murga M, González MC (2015) Origin–destination trips by purpose and time of day inferred from mobile phone data. Transp Res Part C: Emerg Technol 58:240–250

Bayart C, Bonnel P (2018) Mixed-modes survey media and data comparability issues: a French case study. Transp Res Procedia 32:351–362. https://doi.org/10.1016/j.trpro.2018.10.063

Bayart C, Bonnel P, Havet N (2018b) Daily (im)mobility behaviours in France: an application of hurdle models. Transportation

Bissell D, Adey P, Laurier E (2011) Introduction to the Special Issue on Geographies of the Passenger. J Transp Geogr 19(5):1007–1009. https://doi.org/10.1016/j.jtrangeo.2011.06.002

Bissell D (2018) Transit life: how commuting is transforming our cities. In: Transit life: how commuting is transforming our cities. https://doi.org/10.1080/2325548x.2019.1579568

Blondel VD, Decuyper A, Krings G (2015) A survey of results on mobile phone datasets analysis. EPJ Data Science 4(1):1–55. https://doi.org/10.1140/epjds/s13688-015-0046-0

Boyd D, Crawford K (2012) Critical questions for big data: Provocations for a cultural, technological, and scholarly phenomenon. Inf Commun Soc 15(5):662–679. https://doi.org/10.1080/1369118X.2012.678878

Brabham DC (2009) Crowdsourcing the public participation process for planning projects. Plann Theory 8(3):242–262

Büscher M, Urry J (2009) Mobile methods and the empirical. Eur J Soc Theory 12(1):99–116. https://doi.org/10.1177/1368431008099642

Büscher M, Urry J, Witchger K (2010) Mobile methods. Mobile Methods. https://doi.org/10.4324/9780203879900

Calabrese F, Diao M, Di Lorenzo G, Ferreira J Jr, Ratti C (2013) Understanding individual mobility patterns from urban sensing data: a mobile phone trace example. Transp Res Part C 26:301–313. https://doi.org/10.1016/j.trc.2012.09.009

Calabrese F, Di Lorenzo G, Ratti C (2010) Human mobility prediction based on individual and collective geographical preferences. In: IEEE conference on intelligent transportation systems, proceedings. ITSC, pp 312–317. https://doi.org/10.1109/ITSC.2010.5625119

Chen C, Jingtao M, Susilo Y, Liu Y, Wang M (2016) The promises of big data and small data for travel behavior (aka human mobility) analysis. Transp Res Part C 68:85–299

Concilio G, Pucci P, Vecchio G, Lanza G (2019) Big data and policy making: between real time management and the experimental dimension of policies, in: Computational science and Its applications–ICCSA 2019, Part II, Springer, pp 191–202, ISBN 978-3-030-24295-, https://doi.org/10.1007/978-3-030-24296-1_17

D'Andrea A (2006) Neo-nomadism: a theory of post-identitarian mobility in the global age. Mobilities 1(1):95119. https://doi.org/10.1080/17450100500489148

Deloitte (2019) Global mobile consumer 2019, Italy. Available at https://www2.deloitte.com/it/it/pages/technology-media-andtelecommunications/articles/global-mobile-consumer-survey-2019---deloitte-italy---tmt.html. Accessed 11 Apr 2022

DeLyser D, Sui D (2012) Crossing the qualitative quantitative divide II: inventive approaches to big data, mobile methods, and rhythm analysis. Prog Hum Geogr (37):293–305

Docherty I, Marsden G, Anable J (2018) The governance of smart mobility. Transp Res Part A: Policy Pract 115:114–125. https://doi.org/10.1016/j.tra.2017.09.012

Fol S (2009) La mobilité des pauvres. Belin, Paris

De Gennaro M, Paffumi E, Martini G (2016) Big data for supporting low carbon road transport policies in Europe: applications, challenges and opportunities. Big Data Res 6:11–25

Goodchild MF (2007) Citizens as sensors: the world of volunteered geography. GeoJournal 69:211–221

Hubert JP, Armoogum J, Axhausen K, Madre JL (2008) Immobility and mobility seen through trip-based versus time-use surveys. Transp Rev 28(5):641–658. https://doi.org/10.1080/01441640801965722

Jiang S, Ferreira J, Gonzalez MC (2016) Activity-based human mobility patterns inferred from mobile phone data: a case study of Singapore. IEEE Transactions on Big Data 3(2):208–219. https://doi.org/10.1109/tbdata.2016.2631141

Kitchin R (2013) Big data and human geography: opportunities, challenges and risks. Dialogues Hum Geogr 3(3):262–267

Kitchin R (2014) The real-time city? Big data and smart urbanism. GeoJournal 79:1–14

Kourtit K, Nijkamp P, Steenbruggen J (2016) The significance of digital data systems for smart city policy. Socio Econ Plan Sci

Kwan MP (2016) Algorithmic geographies: big data, algorithmic uncertainty, and the production of geographic knowledge. Ann Am Assoc Geogr 106(2):274–282

Lanza G, Pucci P, Carboni L, Vendemmia B (2022) Impacts of the Covid-19 pandemic in inner areas. Tema. J Land Use Mobil Environ 73–89. https://doi.org/10.6092/1970-9870/8915

Lanza G (2021) Data related ecosystems in policy making. The PoliVisu contexts. In: Concilio G, Pucci P, Raes L (eds) The data shake. Opportunities and obstacles for urban policy making. Polimi SpringerBriefs, Springer

Lucas J-F (2020) Digital approaches and mobilities in the big data era. In Jensen OB, Lassen C, Kaufmann V, Freudendal-Pedersen M, Gotzsche Lange IS (Eds) Handbook of urban mobilities, 1st edn. Routledge

Lucas K, Madre JL (2018) Workshop synthesis: dealing with immobility and survey non-response. Transp Res Procedia 32:260–267. https://doi.org/10.1016/j.trpro.2018.10.048

Madre JL, Axhausen KW, Brög W (2007) Immobility in travel diary surveys. Transportation 34(1):107–128. https://doi.org/10.1007/s11116-006-9105-5

Manfredini F, Lanza G, Curci F (2022) Mobile phone traffic data for territorial research. Tema. J Land Use Mobil Environ 2

Manfredini F, Lanza G, Curci F (2023) Mobile phone traffic data for territorial research. Tema. J Land Use Mobil Environ 9–23. https://doi.org/10.6092/1970-9870/8892

Mcneely CL (2014) The big (data) bang: policy. Prospects Challenges 31(4):304–310

Merriman P (2014) Rethinking mobile methods. October, 37–41. https://doi.org/10.1080/17450101.2013.784540

Milbourne P, Kitchen L (2014) Rural mobilities: connecting movement and fixity in rural places. J Rural Stud 34:326–336. https://doi.org/10.1016/j.jrurstud.2014.01.004

Miller HJ (2010) The data avalanche is here. Shouldn't we be digging? J Regional Sci 50:181201. https://doi.org/10.1111/j.1467-9787.2009.00641.x

Milne D, Watling D (2019) Big data and understanding change in the context of planning transport systems. J Transp Geogr 76(February):235–244. https://doi.org/10.1016/j.jtrangeo.2017.11.004

Mokhtarian PL, Salomon I (2001) How derived is the demand for travel? Some conceptual and measurement considerations. Transp Res Part A: Policy Pract 35(8):695–719. https://doi.org/10.1016/S0965-8564(00)00013-6

Motte-Baumvol B, Bonin O (2018) The spatial dimensions of immobility in France. Transportation 45(5):1231–1247. https://doi.org/10.1007/s11116-017-9763-5

Motte-Baumvol B, Nassi CD (2012) Immobility in Rio de Janeiro, beyond poverty. J Transp Geogr 24:67–76. https://doi.org/10.1016/j.jtrangeo.2012.06.012

Motte-Baumvol B, Bonin O, David Nassi C, Belton-Chevallier L (2015) Barriers and (im)mobility in Rio de Janeiro. Urban Studies 53(14):2956–2972. https://doi.org/10.1177/0042098015603290

Motte-Baumvol B, Bonin O, David Nassi C, Belton-Chevallier L (2016) Barriers and (im)mobility in Rio de Janeiro. Urban Studies 53(14):2956–2972.https://doi.org/10.1177/0042098015603290

Ni L, Wang X (Cara), Chen X (Michael) (2018) A spatial econometric model for travel flow analysis and real-world applications with massive mobile phone data. Transp Res Part C: Emerg Technol 86(April 2017):510–526. https://doi.org/10.1016/j.trc.2017.12.002

Picornell M, Ruiz T, Lenormand M, Ramasco JJ, Dubernet T, Frías-Martínez E (2015) Exploring the potential of phone call data to characterize the relationship between social network and travel behavior. Transportation 42(4):647–668. https://doi.org/10.1007/s11116-015-9594-1

Pucci P, Vecchio G (2019a) Enabling mobilities. PoliMi Springer Brief. Springer

Pucci P, Vecchio G (2019b) Trespassing for mobilities. Operational directions for addressing mobile lives. J Transp Geogr 91. https://doi.org/10.1016/j.jtrangeo.2019.102536

Pucci P, Gargiulo C, Manfredini F, Carpentieri G (2022) Mobile phone data for exploring spatio-temporal transformations in contemporary territories. Directory Open Access J. https://doi.org/10.6093/1970-9870/9534

Pucci P, Manfredini F, Tagliolato P (2015) Mapping urban practices through mobile phone data, PoliMI SpringerBriefs Series, ISBN 9783319148328; WOS:000363269000009

Rabari C, Storper M (2015) The digital skin of cities: urban theory and research in the age of the sensored and metered city, ubiquitous computing and big data. Camb J Reg Econ Soc 8:27–42

Richard O, Rabaud M (2018) French household travel survey: The next generation. Transp Res Procedia 32:383–393. https://doi.org/10.1016/j.trpro.2018.10.065

Salat H, Smoreda Z, Schläpfer M (2019) Mobile phone data's potential for informing infrastructure planning in developing countries. http://arxiv.org/abs/1907.04812

Salkind NJ (2007) Encyclopedia of measurement and statistics. SAGE, Thousand Oaks

Schwanen T (2015) Beyond instrument: smartphone app and sustainable mobility. Eur J Transp Infrastruct Res 15(4):675–690

Schwanen T (2016) Geographies of transport I: reinventing a field? Prog Hum Geogr 40(1):126–137. https://doi.org/10.1177/0309132514565725

Schwanen T (2017) Geographies of transport II: eeconciling the general and the particular. Prog Hum Geogr 41(3):355–364. https://doi.org/10.1177/0309132516628259

Semanjski I, Bellens R, Gautama S, Witlox F (2016) Integrating big data into a sustainable mobility policy 2.0 planning support system. Sustainability 8:1142

Shaw J, Hesse M (2010) Transport, geography and the 'new' mobilities. Trans Inst Br Geogr New Ser 35(3):305–312

Shaw J, Sidaway JD (2010) Making links: on (re)engaging with transport and transport geography. Prog Hum Geogr 35(4):502–520

Sheller M (2014) The new mobilities paradigm for a live sociology. Curr Sociol 62(6):789–811. https://doi.org/10.1177/0011392114533211

Sikder S, Pinjari AR (2012) Immobility levels and mobility preferences of the elderly in the United States: evidence from 2009 national household travel survey. Transp Res Rec 2318:137–147. https://doi.org/10.3141/2318-16

SNAI Inner area report (2018–2020)—Appennino smart. Adattamento intelligente per cambiare gli schemi d'azione e superare le criticità con nuove idee. Available at https://www.agenziaco esione.gov.it/wpcontent/uploads/2020/07/Definitivo_Strategia_AI_Appennino_PC_PR_dic_18.pdf. Accessed 11 Apr 2022

Stopher P, Greaves S (2007) Household travel surveys: where are we going? Transp Res Part A 41:367–381

Thakuriah P (Vonu), Tilahun NY, Zellner M (2017) Big data and urban informatics: Innovations and challenges to urban planning and knowledge discovery. Springer Geography. https://doi.org/10.1007/978-3-319-40902-3_2

Tranos E, Mack E (2019) Big data: a new opportunity for transport geography? J Transp Geogr 76(August 2018), 232–234. https://doi.org/10.1016/j.jtrangeo.2018.08.003

Tu W, Cao J, Yue Y, Shaw SL, Zhou M, Wang Z, Chang X, Xu Y, Li Q (2017) Coupling mobile phone and social media data: a new approach to understanding urban functions and diurnal patterns. Int J Geogr Inf Sci 31(12):2331–2358. https://doi.org/10.1080/13658816.2017.1356464

Urbanek A (2019) Data-driven transport policy in cities: a literature review and implications for future developments. In: Advances in intelligent systems and computing, vol. 844. Springer International Publishing. https://doi.org/10.1007/978-3-319-99477-2_6

van Wee B, Annema J, Banister D (eds) (2013) The transport system and transport policy. Elgar, Cheltenham

Vannini P, Scott N (2020) Mobile ethnographies of the city. In: Jensen OB, Lassen C, Kaufmann V, Freudendal-Pedersen M, Gotzsche Lange IS (eds) Handbook of urban mobilities, 1st edn. Routledge

Vecchio G, Tricarico L (2019) "May the force move you": roles and actors of information sharing devices in urban mobility. Cities 88(November 2018):261–268. https://doi.org/10.1016/j.cities.2018.11.007

Vendemmia B, Pucci P, Beria P (2021) An institutional periphery in discussion. Rethinking the inner areas in Italy through the lens of accessibility. Appl Geogr 135:102537

van Wee B (2013) Land use and transport. In: van Wee B, Annema JA, Banister D (eds) The transport system and transport policy. Elgar, Cheltenham

Welch TF, Widita A (2019) Big data in public transportation: a review of sources and methods. Transp Rev 39(6):795–818

Williams NE, Thomas TA, Dunbar M, Eagle N, Dobra A (2015) Measures of human mobility using mobile phone records enhanced with GIS data. PLoS ONE 10(7):e0133630. https://doi.org/10.1371/journal.pone.0133630

Chapter 4
Assessing Spatial Conditions Enabling Immobility

Abstract This chapter examines the spatial and social conditions enabling chosen, reversible immobility, framed through the concept of accessibility by proximity. This concept informs integrated land use and transport policies that promote access to essential services within close physical proximity, primarily via active mobility, thereby reducing the need for long-distance travel. Measuring accessibility by proximity is essential for identifying territorial gaps in service provision and understanding when immobility can be configured as a choice or a constraint, as outlined in Chap. 2. Even if extensively researched, measuring accessibility by proximity is a challenging analytical effort, particularly in medium- to low-density areas, where the concept of proximity should be reconsidered compared to the most common approaches usually applied in high-density urban cores. The chapter's empirical section addresses these challenges by applying a methodology for measuring accessibility by proximity in a low-density context, offering new directions for analyzing accessibility in peri-urban areas.

4.1 Introduction

The analysis of realized mobility presented in the previous chapter of the book allows for identifying possible differentials in terms of frequency and spatial extension of movements and revealing forms of relative immobility that may characterize some socio-spatial contexts more than others. The analysis of actual mobility practices thus offers a snapshot of the differential dynamics at work. However, it does not create the conditions for an in-depth study of the social and spatial factors that can at least partially explain and qualify these differentials in terms of potential social inclusion and participation.

As proposed in the conceptual framework illustrated in Chap. 2, the analysis and measurement of accessibility by proximity can contribute to qualifying the potential social and spatial implications of immobility and classifying it as either reversible or constrained. This proposal is based on the conception that accessibility by proximity expresses the individual potential for activity participation, interaction and exchange

© The Author(s), under exclusive license to Springer Nature Switzerland AG 2025
G. Lanza, *Enabling Immobilities: Social and Spatial Implications for Urban Planning*,
PoliMI SpringerBriefs, https://doi.org/10.1007/978-3-031-80999-6_4

based on the conditions that a person can achieve as provided by locations, infrastructures, mobility services, as well as new forms of virtual connection (Martens 2016) in proximity. From this perspective, accessibility by proximity becomes a promising concept for integrating multi-sectoral policies, as it simultaneously considers aspects related to transport, land use, and the spatial provision of services and opportunities, acknowledging the plurality of mobility and access experiences expressed by individuals in different spatial and social contexts. In planning terms, the concept of accessibility by proximity can thus guide specific integrated land use and transport policies aimed at improving access to valued daily activities and opportunities via low mobility (or relative immobility).

However, for these policies to be effectively designed, implemented, and targeted to the appropriate social and spatial contexts—an aspect that is at the same time complex and highly debated in planning, as will be discussed more thoroughly in the following sections—it is necessary to employ robust assessment methodologies and tools to detect and quantify current differences in access to opportunities in proximity. By assessing in which social and spatial contexts immobility can be traced back to a choice or a constraint and orienting policy measures to increase low levels of accessibility by proximity and enable chosen, reversible immobility, in accordance with the possibilities offered by the context of analysis, these assessment tools can significantly contribute to plan for more accessible, sustainable and fair cities.

Among the four fundamental issues regarding the translation of the concept of immobility into urban planning identified in Chap. 2, here the focus is set on the second point which concerns exploring the implications of immobility and the spatial conditions that can support reversible immobility through the measurement of accessibility by proximity. As already clarified in Chap. 2, while recognizing the importance of connectivity in ensuring indirect virtual access to goods and resources (Levine et al. 2019), the focus here will be on accessibility by proximity intended in physical terms, not directly including considerations related to digital connectivity.

The chapter opens with an account of the main challenges related to accessibility by proximity measurements, identifying some critical points that concern, in particular, the definition of activities to be included in the measurement, the transport modes considered, and the user profiles included in the simulation (Sect. 4.2). The analysis of the limits and opportunities identified forms the basis of an innovative experiment in measuring accessibility by proximity in the context of the Piacenza Apennines, presented in Sect. 4.3 as an example of an evaluation conducted in an area characterized by specific complexities related to the harsh orography, the socio-demographic trends affecting the territory and the spatial availability and distribution of local services. In Sect. 4.4, the results of the empirical experiment are consequently discussed and illustratively compared with the results of the immobility analysis presented in Chap. 3. Thus, relevant theoretical and operational insights useful for the progressive evolution and improvement of these essential territorial analysis tools are identified. Conclusive remarks close the chapter.

4.2 Methodologies and Challenges for Accessibility by Proximity Measurement

The measurement of accessibility by proximity is situated within the planning literature which has long recognized the role of accessibility as an enabler of individual participation in social opportunities and networks (Geurs and van Wee 2004; Farrington and Farrington, 2005; Preston and Rajé 2007; Currie and Delbosc 2011; Lucas 2012; Martens 2017; Pucci and Vecchio 2019) and its measurement as a way to detect different forms and degrees of social exclusion *"due in whole or in part to insufficient mobility in a society and environment built around the assumption of high mobility"* (Kenyon et al. 2002: 210–211). Accessibility by proximity—and its measurement—can be interpreted as a specific declination of the more general concept of accessibility understood as the ability to participate in valued spatially-distributed activities. Indeed, accessibility by proximity specifically focuses on the conditions of access to activities that are relevant or essential for daily life and should, therefore, be available within the physical proximity of individuals to limit the need for travel and ensure well-being and social inclusion. Additionally, metrics of accessibility by proximity should primarily focus on active and collective modes of transport to reach such services, integrating possible elements regarding the direct interactions and experience of the spaces that people may have while moving actively.

Metrics of accessibility and, more specifically, accessibility by proximity serve as social indicators that allow identifying relative differentials in access to opportunities among social groups or places that can be expressions of existing inequalities and potential social exclusion. In this perspective, accessibility assumes a normative value since its measurement becomes a people-focused and needs-based key policy indicator for decision-makers, useful for assessing how land-use patterns, transport systems, and policy plans may affect the travel capacity of populations characterized by different identities, incomes, skills, and ability levels (Farrington 2007; Benenson et al. 2010, 2017; van Wee et al. 2013; Niedzielski and Boschmann 2014; Guzman et al. 2017; Guzman and Oviedo 2018; Martens 2016, 2017). Accessibility then becomes part of the project of explicitly integrating the notion of space into the understanding of social justice and designing policies to address it while considering how the characteristics of individuals, groups, and communities affected by low levels of accessibility may become potential barriers to participation (Farrington and Farrington 2005; Kamruzzaman et al. 2016).

However, designing effective metrics is a challenging endeavor due to the complexity and variability of the conditions to be considered. According to the well-known schematization proposed by Geurs and van Wee (2004) on the components of accessibility, a rigorous metric should simultaneously take into account four relevant components: the spatial distribution of supply and demand for available activities and opportunities (land use component), the efficiency of transport systems in terms of travel time, cost, and effort (transport component), the time constraints that may affect the availability and reachability of spatial opportunities (temporal component), and finally, the needs, opportunities, and abilities that influence one's possibilities to translate mobility potential into action (individual component). Given the complexity of a measurement that considers all these dimensions simultaneously, it is clear why accessibility—and specifically accessibility by proximity metrics—have a still limited role in integrated land use and transport planning practice (Boisjoly and El-Geneidy 2017a). This limited use is mainly related to the complexity of the concept and the manifold components that should be measured and included to provide a theoretically sound measure. Additionally, accessibility by proximity metrics face several issues that create opportunities for debate between different perspectives and require specific reflections and considerations, partially discussed in Lanza et al. (2023).

The first consideration regards the fact that any accessibility metric requires a set of activities to be included in the assessment as the "destinations" to be accessed. The definition of the set of activities in an accessibility by proximity metric should consider the fact that accessibility is a relativistic and contingent concept, making the participation in activities and evaluating its levels to be defined according to the culture and needs of each place, social group, and individual. According to Farrington (2007), relativism is unavoidable and encourages a reflection on how different values might be spatially and culturally reflected into accessibility needs in different ways in different societies. Considering the extreme variety of possible needs and desires of individuals, as well as the objective impossibility and economic and environmental unsustainability of providing extensive access to every place, some authors (Martens et al. 2014; Lucas et al. 2016) have proposed to prioritize the selection of activities focusing on those that, in the context of analysis, play a basic role in guaranteeing social inclusion and in allowing activity participation, leading to the pursuit of basic accessibility (Vecchio 2019, p. 24). The concept of basic accessibility is related to the principle of sufficientarianism, a philosophical approach to equity intended as the distribution of social benefits and burdens of a good. In this perspective, accessibility is a good that should be ensured at least to a minimum basic level, expressed by the possibility to access a set of essential activities and services, so that people finding themselves below the sufficient threshold may be exposed to an injustice (van der Veen et al. 2020). The theoretical principle of sufficientarianism establishes that, in a normative process, the objective of transport and land use policy and planning should be to set context-sensitive definitions of essential activities and thresholds through which to detect and respond to accessibility shortfalls focusing primarily on the relatively most disadvantaged populations and territories. Thus, the choice of a set of basic activities to be included in the assessment should be the outcome of deliberative processes (Martens 2017, p. 207), seeking a common ground to guide the approach toward a fair public policy agenda for a community. This approach could be developed through crowd- sourced listening, survey and direct participation systems to identify and assess the relevance of the sets of activities (Lanza et al. 2023)—and the benefit produced—to be considered in the analysis (Büttner et al. 2022), as proposed in few but relevant tools such as the Neighborhood Destination Accessibility Index (NDAI) developed by Witten et al. (2011). Indeed, it is important, at least from a theoretical and operational perspective, to move away from standardized lists of services and functions to be considered in the assessment as those usually proposed in different approaches to accessibility by proximity such as the 15 min city model which defines a generic list of social functions, namely living, working, healthcare/caring, education, commerce, and entertainment as commonly essential and necessary (Moreno et al. 2021). Clearly, these services address a set of needs that are certainly important and "basic", but it would still be crucial to flexibly adapt the list considering the specific needs of the populations in the areas of analysis, so as to adequately respond to their accessibility by proximity needs. In this perspective, the concept of accessibility by proximity could even be considered as a facet of basic accessibility (Lanza and Pucci 2022) in which a context-sensitive planning criterion (and related measurement tools) is adopted based on the availability and spatial proximity to specific and locally-defined essential services and opportunities and the promotion of active and sustainable ways to access them. Ultimately, knowing which services are needed in proximity for daily life by a community and its members is a key element for promoting accessibility in proximity and reversible immobility.

The second point concerns the modes of transport used to reach the selected destinations. Considering that the concept of accessibility by proximity, as highlighted, involves the promotion of forms of active, collective, or shared mobility, related metrics should specifically focus on these types of transport modes. The emphasis on active proximity mobility (e.g., walking or cycling) is particularly significant in the idea that accessibility by proximity can create conditions for reducing long-range travel, generating forms of reversible relative immobility. Simultaneously, considering the role and availability of shared and collective mobility systems in the measurement allows for determining how easy and feasible it is to reach services and activities that are not strictly intended for daily use and which cannot always be guaranteed in everyone's immediate proximity due to the effect of specific economies of scale (e.g., in the case of major functions such as hospitals or secondary education facilities).

However, the assessment of accessibility by proximity through active mobility introduces concrete challenges since it requires investigating the physical and perceptual characteristics of the spatial contexts under examination and their influence on active mobility behaviors. Consequently, the accessibility analysis should be combined with two well-developed and relevant concepts of urban and transportation studies: walkability and cyclability. According to Southworth (2005), *"walkability is the extent to which the built environment supports and encourages walking by providing for pedestrian comfort and safety, connecting people with varied destinations within a reasonable amount of time and effort, and offering visual interest in journeys throughout the network"* (p. 248). Such a definition can apply also for bicycle use. From this perspective, the level of accessibility by proximity will depend not only and exclusively on the distance or time required to reach a particular activity, but also on the possibility of reaching it through routes and spaces that are technically well designed, safe, attractive, and able to promote sociality, especially for individuals moving at slow speed, who are typically more aware and exposed to the environment and its variables than drivers are (Clifton et al. 2007; Lanza et al. 2023). This focus allows overcoming a common limitation of most accessibility measurement tools, which primarily analyze travel by car or public transportation (Pajares et al. 2021; Silva et al. 2017) and requires an in-depth consideration of the interplay between the propensity, capabilities, and attitudes for active mobility modes and the characteristics, both objectively and subjectively perceived, of the surrounding environment (Ewing and Handy 2009).

This set of observations leads to the third consideration, which concerns which profiles should be included and based on which evaluation process of their preferences and capabilities in accessibility assessment as agents of the simulation. Indeed it is possible to conceive different experiences of access in proximity, linked both to the existence of multiple opportunity types in terms of needed services to be accessible (van der Veen et al. 2020) and to the different ways in which the abilities and possibilities of individuals will be confronted on a daily basis with the specific morphological, social, and functional characteristics of the proximity spaces in which they (do not) move. The development of accessibility by proximity metrics, therefore, requires a "local accessibility way of thinking" (Handy 2020) considering the specific needs and desires of different social groups characterized by different physical and cognitive capabilities that also affect their accessibility and mobility attitudes. In doing so, more attention in accessibility-by-proximity-inspired planning strategies should also be paid to the needs of non-standardized population profiles, such as people with mobility impairments (Büttner et al. 2022) unlike those commonly considered in accessibility analysis, whose travel behaviors are generally assumed to be unaffected by the spatial frictions generated by the places in which they move. This is represented, for example, by the use of undifferentiated average standard speeds for calculating catchment areas in many gravity-based accessibility assessments, without including specific impediments related to physical factors such as the presence of barriers, slopes, or perceptual factors such as the sense of insecurity. Indeed, the evaluation and quantification of the costs and benefits associated with various objective and subjective factors of walkability and cyclability (Iacono et al. 2010) should be an important concern within any active mobility-based accessibility measurement. While conducting such assessments at the individual level may be complex and impractical, it is feasible to do so for different social groups that share specific mobility habits and accessibility needs. Several tools in the literature support this approach by differentiating users in terms of modal choice, for instance focusing on cyclists (Lowry et al. 2012; Arellana et al. 2020), the elderly (Gaglione et al. 2022), children (Buck et al. 2011), visually impaired individuals (Campisi et al. 2021), or by adapting established methodologies like the Walkability Index developed by Frank et al. (2010) to the unique needs of impaired pedestrians (Portegijs et al. 2017). The specific experience of different population profiles in active mobility simulations often involve their direct engagement through designed survey methods to identify a hierarchy of walkability and cyclability factors, assigning different weights to elements based on their influence on active mobility attitudes, tailored to various people's abilities, possibilities, needs, and preferences. Although this approach shows promise in designing context- and people-sensitive tools, it is applied by only a limited number of assessment tools, such as the Quality of

Pedestrian Level of Service (QPLOS) by Talavera-García and Soria-Lara (2015) or the Pedestrian Accessibility Tool (PAT) (Erath et al. 2017) mainly due to the resource-intensive nature of direct public involvement.

Beyond the three considerations mentioned above, there are other technical aspects that arise when an accessibility by proximity indicator is operationally implemented and applied in concrete cases. A first point concerns the level of social and spatial aggregation to which the measurement can be referred: the more detailed and extensive the information referring to the complex and varied accessibility needs of individuals, the higher the level of accuracy of the indicator. On the contrary, highly aggregated measures might prevent accounting for the individual complexity and variability of mobility practices and accessibility potentials, overlooking the specificities of individuals and places (Preston and Raje 2007, Li et al. 2011; Vecchio 2019). However, it remains important to note, as done by Geurs and van Wee (2004), that a person-based perspective requires information and data that are often unavailable and high computational capacity, directing analyses toward the listening of small study groups and their accessibility needs through qualitative methodologies of analysis and research, thus limiting the spatial representativeness of the indicator.

A second relevant aspect concerns the geographical and social scale of analysis. It is possible to hypothesize a trade-off between the possibility of obtaining socially and spatially disaggregated results and the extent of the area of study. Measurements on very geographically limited micro-regions that considers the local social make-up and the diverse needs and capabilities of different local populations can return the level of accessibility to activities and opportunities even at the individual scale and household level with higher detail. In contrast, at the mesoscale, indicators may show the potential for movement at the neighborhood level with a varying degree of social disaggregation based on the availability of data and the possibility to provide qualitative need assessment of the local population to expand and contextualize the set of basic activities included in the analysis. Finally, macro measurements on extended contexts are likely used in strategic terms to identify, mainly through location and transport-based measurement, the degree to which land use and transport network facilitate travel from one area to another for aggregated population groups (Jones and Lucas 2012).

A third and final aspect concerns temporality, namely the possibility of dynamically measuring accessibility by taking into account how it varies considering time as a variable. To this extent, accessibility can be influenced by the hourly availability of the opportunities to be reached, the existence of time constraints that may affect our travel activity, the manifestation of circumstances independent from our will but which may nonetheless affect our level of access to activities and opportunities such as traffic congestion, which is often a byproduct of accessibility (Mondschein and Taylor 2017). Considering the time component as fixed may lead to deceptive or biased results (Farber et al. 2014; Järv et al. 2014, 2018, Garcia-Albertos et al. 2019).

Taking all the previously described dimensions simultaneously into account is crucial to discover the causes and the effects of differentials in the levels of physical and virtual accessibility to urban opportunities as experienced by different individuals and groups and to shape inclusive mobility policies accordingly.

However, the lack of reliable and usable quantitative and qualitative data at the aggregated and individual levels and the significant differences in data availability in different global contexts represents substantial constraints in promoting an extensive use of accessibility measures for policy-making purposes. These difficulties demonstrate why accessibility measures have been mostly experimented with within the scientific field rather than professional planning (te Brömmelstroet 2010; Boisjoly and El-Geneidy 2017b) and, when accessibility measurement tools are routinely used for transport planning (e.g., PTAL index developed by Transport for London), they often present certain limitations which include not fully addressing the considerations illustrated in this section, such as the diversification of profiles, the contextual selection of services included, and the attention to specific conditions that influence walkability and cyclability. In addition to these, there is also the so-called rigor-relevance dilemma: the higher the rigor and precision one intends to achieve in making the measurement tool, the lower its practical relevance due to the complexity of use by end-users (Papa et al. 2016).

In conclusion, accessibility by proximity measurement is a suitable and relevant activity—yet still challenging in terms of operationalization—to identify the socio-spatial extent of inequalities that could signal potential disadvantages in participation and social inclusion for certain individuals and territories, especially when low levels of accessibility are coupled with conditions of relative immobility. Conversely, the measurement of accessibility can also reveal situations where physical proximity to necessary and desired opportunities can replace the need for high mobility by creating conditions for relative, reversible immobility induced by the socio-spatial configurations of settlements that meet the inhabitants' needs.

All these conditions can be analyzed with a sufficiently complete and detailed dataset regarding the spatial characteristics of a given context, the preferences and possibilities related to active mobility among different population profiles, and by developing appropriate methodologies that take into account the issues and considerations discussed in this section. As will be seen in the next part, various aspects have been considered in constructing an experimental accessibility by proximity index inspired by the IAPI index (Lanza et al. 2023) and applied in the context of the Piacenza Apennines.

4.3 An Experimental Approach to Accessibility by Proximity Measurement

In this chapter, a method for measuring accessibility by proximity is presented, focusing on a challenging territory, namely the Apennine area of the Piacenza Province. This area is particularly interesting from the perspective of developing accessibility by proximity metrics because it exhibits, in addition to the specific

socio-demographic dynamics illustrated in the previous chapter, further characteristics related to the availability and accessibility of daily services and amenities,
which are common to many medium-to-low population density peri-urban contexts.
Addressing accessibility by proximity in these contexts introduces theoretical and
practical questions, as the concept is typically associated to dense urban settings.
This raises the question of whether ideas of a just and sustainable society proper of
specific visions and urban models such as the 15 min city can also apply to non-urban
areas, such as suburbs and rural communities (Poorthuis and Zook 2023).

For this reason, it will be necessary to provide a measurement tool adapted to a
context where, outside the two main centers of Bobbio and Bettola, located in the two
major valleys (Valle Trebbia and Valle Nura), the supply of everyday services tends to
be minimal, if not entirely absent. Moreover, even when facilities are present locally,
their availability in smaller municipalities is often limited by restricted opening hours
that do not guarantee continuity of operation or are based on seasonality, in an area
that sees several touristic arrivals mostly concentrated in the summer (Lanza et al.
2022). The problem of the low availability of services and amenities in a morpholog-
ically complex territory with low settlement density generates the need—for those
inhabitants who can—to travel long distances and spend significant time reaching
the first available opportunities, often essential for their sustenance and daily life. At
the same time, the aging process and the exodus of younger populations have led to
the closure or centralization of many activities and services, as in many other rural
areas. Today, access to these services is often inadequate, resulting in processes of
territorial and socio-economic marginalization that simultaneously cause and result
from scarce accessibility, undermining the livability of rural communities (Vitale
Brovarone and Cotella 2020).

The long distances and the rough topography suggest that most travel should be
car- or bus-based, while active mobility options are significant only for proximity
trips within individual centers. Indeed, the public transport system covers the territory
relatively well, with a network designed to follow the valley structure. However, due
to low density and travel demand, lines serving remote hamlets usually have low
frequency and are partly operated under a Demand Responsive Transport (DRT)
scheme. A reliable and relatively frequent service is only available along the main
roads to and from the major cities in the flat area of the province.

Under these conditions, the daily activities of sustenance and access to essential
services by those living in marginal settings are severely limited for those who do not
have a private means of transport available. In this scenario, mobility by public transport can be largely marginal (Beria 2020) and determine an imbalance, in terms of
activity participation, between motorized and non-motorized individuals, as well as
creating the conditions for the development of forms of car dependence. As discussed
in Chap. 2, the ability to access a private means of transport has emerged as a strong
determinant of people's propensity to move, particularly valid in a context of low
population density and opportunity. Moreover, spatial conditions explain the high
rate of car ownership per inhabitant which is above national averages (ACI 2018).
These figures suggest that lack of access to a car as an owner or passenger due to

dispersed land-use patterns and lack of accessible services, poor public transport and the demographic characteristics of the population may generate situations of spatial immobility and undermine social inclusion due to transport disadvantage (Shergold et al. 2012; Mattioli 2014).

Simultaneously, owning a car should not necessarily be considered a wealth index. Forced car ownership, induced by the need to rely on car-based mobility to ensure participation and access to essential services, can generate very high economic and environmental costs and represent a burden for individuals when cheaper options are unavailable (Currie and Delbosc 2011; Mattioli 2014), potentially exacerbating the already significant outmigration from peripheral areas.

In this context, an accessibility by proximity index was applied, specifically designed based on the methodology proposed by Lanza et al. (2023) with their Inclusive Accessibility by Proximity Index (IAPI) but adapted to a structurally different context characterized by a significant scarcity and poor update of detailed and comprehensive geospatial data, both crowdsourced (e.g., retrievable from Open Street Map) or produced by local authorities.

The calculation was conducted considering different transport modes that reflect the main possibilities available in a mountainous context, namely walking, public transport, and car usage, to obtain a comparative result that identifies the gaps between the opportunities offered by different mobility options. The proposed hypothesis is that measuring accessibility by proximity at municipal and submunicipal levels— thus, considering the relations existing between the central core of municipalities and the several hamlets dispersed within the municipal administrative boundaries—can be a useful tool, from a normative perspective, as it allows for the identification of those territorial contexts inhabited by people who can live in areas offering them the possibility to express their reversible immobility thanks to the presence and avail-ability of functions or who, conversely, directly experience limited physical access to some basic daily activities. This can inform targeted and selective interventions to address emerging forms of accessibility and mobility-related marginalities.

The accessibility by proximity index has been constructed to evaluate simultane-ously the availability and the level of spatial–temporal accessibility to basic services within each municipality of the study area. The index is first calculated at the sub-municipal level and then recalculated at the municipal scale, thus making it possible to directly compare the results of the accessibility index with the levels of realized daily immobility recorded through TIM data presented in Chap. 3, as deepened in the next section.

The delimitation of the measure of accessibility to municipal boundaries is an arbitrarily methodological choice: in a territory administratively divided into large municipalities with dispersed hamlets, it is likely that a particular service available in another municipality could be more easily or preferably reachable than the one located within the municipal boundary. However, the purpose of the index is to identify (i) those territorial conditions in which the use of a specific service not avail-able within the municipality requires moving to another center (*service availability* dimension)—namely, contexts where a condition of prevalent relative daily immo-bility may create limits to activity participation and (ii) the different conditions of accessibility in terms of spatial–temporal distance to the resources available within the same municipality (*service accessibility* dimension). This last information is helpful for identifying which sub-areas—and which inhabitants—potentially experience conditions of relative disadvantage in the use of proximity services that may be available in their municipality, especially if they are unable to move independently to reach them.

The proposed index simultaneously considers both dimensions to describe some circumstances in which conditions of immobility can or cannot be traced back to a problem, as well as identifying situations in which being mobile over long distances (both within the same municipality and towards other areas of the territory) represents a potential need, but also a cost that the inhabitants assume.

From an operational point of view, the proposed measure of accessibility takes its cue from the Hansen's gravitational model (1959):

$$Ai = \sum a\bar{J}f(d_{IJ})$$

where, in the proposed definition, Ai equals the accessibility of zone i (origin zone within a municipality), aj the attractiveness of j (locations of basic services within the municipality), and f (dij) the function of the distance between i and j.

At the municipal scale, the measure describes the level of potential accessibility to each available service from each populated settlement. The essential services (j) to be included in the calculation were chosen based on the SNAI approach, which estimates the level of potential marginality of a center based on the spatial and temporal distance from the first provider pole of advanced healthcare, higher education and rail services. However, in this work, the list of services was reshaped and expanded with other resources of local relevance that may be important on a day-to-day basis from a proximity perspective in the study context. In particular, the process of reshaping the list of services took into consideration the results of the analysis conducted during the design phase of the SNAI local strategy for the Piacenza Apennine area (SNAI Inner area report 2018–2020) to identify some deficiencies concerning services and opportunities perceived by the populations of which the mayors involved in the SNAI roundtables have become spokespeople. These deficiencies mainly regard the limited diffusion of non-specialized health services of proximity (pharmacies, GP), low accessibility to first and second-grade schools, lack of commercial activities of subsistence, and absence of cultural and recreational opportunities. From these results, a list was constructed (Table 4.1) that represents the first attempt, clearly not exhaustive, to identify basic services to be included in the index calculation.

The services selected in this pilot were defined based on stakeholder engagement processes that identified them as lacking within the Apennine area but essential to local communities. However, the available information did not allow for the establishment of criteria by which to weigh services according to their relevance to the local population. This means that each service considered assumes equal importance in the economy of the measurement, although significant differences may exist in

Table 4.1 List of included services

Typology	Service
Education	Nursery, Kindergarten, Primary, Middle and High schools
Healthcare	GP, Pharmacy, Casa della salute, Hospital
Commerces	Grocery store, Post office, Bank
Leisure	Cinema, Theatre, Sport field, Swimming pool

the needs and desires of the different social and demographic groups that inhabit the area. As will be suggested in the next chapter, a population engagement approach may create the conditions for refining these evaluation tools by upgrading the basket of services and assessing their relevance to the daily lives of local inhabitants.

The areas of origin of the displacement (i) towards services (j) have been identified using the geometry of the census tracts, minimum territorial units for the collection of statistical data, selecting only the inhabited settlements according to the most complete available survey (ISTAT 2011). The function of the distance between the centroid of each settlement and punctual services within the same municipality f (dij) is expressed in terms of time—distance by defining the areas accessible from each service by calculating isochrones of 5, 10, and 15 min by foot, car and public transport. Other means of transportation (such as bicycles) are not considered in the calculation, mainly because of the unfavorable terrain of the area that limits their usability. The time thresholds have been set considering that, at the pedestrian level, the dimensions of the inhabited centers are usually small and such that walking times of 15 min are sufficient to cross the settlements and reach the areas where essential services are located. The same thresholds have been applied to public transport and cars, considering the same logic adopted in the case of pedestrian movements but recalibrating it on the size of the municipalities and the speed of motorized vehicles. For all three modes of transport, a travel time from the area of residence to the nearest service of less than 5 min indicates a high level of accessibility, medium if the times are between 5 and 10 min, low if between 10 and 15, and very low if over 15 min. Other values have been applied, i.e. following Carlos Moreno's proposal, which extends the thresholds to 30 min in less densely populated areas according to the 30 min territory concept. This approach extends the idea that accessibility by proximity may require a certain level of mobility in low-density contexts, placing more emphasis on public transportation or sharing solutions (Moreno et al. 2021; Poorthuis and Zook 2023). However, since the analysis proposed in this work focuses directly on the scale of the individual municipality and does not define a broader territorial perspective, it was deemed useful to maintain the 15 min threshold. In any case, this assumption still foresees that the principle of accessibility by proximity should be extended over larger scales and considering different modes of transport other than pure active mobility in these territories, particularly if proximity can be ensured, in most of the cases, through physical travel—even if relatively short in terms of time and distance—by car or public transport.

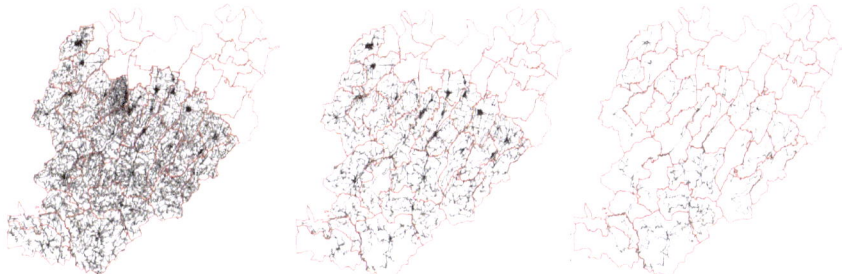

Fig. 4.1 Pedestrian graph (top left), car graph (top right), public transport graph (bottom centre)

The calculation of the isochrones has been carried out from every single service facility using as a simulation base three detailed digital graphs (Fig. 4.1) of the province's road network downloaded from OSM (Open Street Map) using the Osmnx Python Script (Boeing 2017).

Each graph contains information regarding the modes of transport available in the territory: the pedestrian graph contains the geometry of all the roads (arcs) that can be publicly accessible by walking; the car graph features only the geometry of the roads where the transit of vehicles is allowed; the public transport graph considers the effective paths of the bus lines serving the area extracted from the local GTFS (general transit feed specification) released by the local transit authority (SETA and Tempi Agenzia). Each of the three resulting graphs requires an ad hoc construction. In the pedestrian graph, a travel speed along the arcs of the network of 4 km/h was assumed, considering the gaits of people of different ages (Pinna and Murrau 2018). Some impedance factors related to the type of routes have also been set to simulate the slowing effect generated by stairs or unpaved paths present in the Apennine area, where the speed is therefore halved to 2 km/h. In the car graph, travel speeds are attributed to the arcs according to the hierarchy of the road network and the speed limits imposed on-site according to the following criteria: Main suburban roads: 70 km/h; Urban roads: 50 km/h; Suburban mountain roads—urban roads at reduced speed: 30 km/h. An impedance factor is applied based on direct driving experience on Apennine roads. In the case of the public transport graph, speeds have been attributed to the arcs of the network covered by the bus lines and quantified considering the performance arising from the official service timetables declared by the local transit authority in the GTFS. Since the bus network does not reach all the inhabited settlements, the calculation assumes as origins and destinations only the locations and service facilities located within a five-minute walk calculated on a pedestrian graph from the nearest local bus stop. Moreover, an impedance factor linked to the frequency of the service has been introduced, following the methodology proposed by Pucci et al. (2022a). The level of space–time accessibility from a bus stop to a specific location is summed to the frequency of the lines serving the stop to express the discomfort due to the low reliability and frequency of passage characterizing peripheral lines in the area. As described, each of the three graphs was characterized by introducing some impedance factors to the arcs of the network related to the quality of routes and transit services. These factors aim to simulate a subjective response to some objective conditions of the space, for example identifying elements that may represent obstacles for people with reduced mobility (as proposed in the pedestrian graph), considering the reliability of different public transport lines in terms of hourly frequency, or taking into account the difficulty of driving on mountain roads. As proposed by Pucci et al. (2022b) and Lanza et al. (2023), the characterization of the graph could be further refined through more accurate data on the quality of paths and the perception of space and transport services by local users collected through interviews, questionnaires or digital tools. This refinement would be based on the mapping of a series of impedance factors, both positive (with a limiting effect in terms of speed and extension of the catchment area of a service measured through the isochrones) and negative (to simulate the impact of factors that instead facilitate travel by making it more comfortable). Moreover, impedance factors could be defined and weighted into the simulation graph based on the potential impact on different populations characterized by unequal movement capabilities (e.g., people with reduced mobility) so as to simulate different mobility and access experiences to basic services in the area of analysis. Nevertheless, in this experimental phase, only some basic impedance factors have been assumed in the characterization of the graph both due to the limited availability of data in the context of investigation (in rural areas, spatial data available on OSM tend to be less detailed than in dense urban areas) and to the necessity to further explore the perceptual dimension of space by the different users who travel across it in order to achieve an accurate measurement.

After concluding the preliminary construction of the graphs and mapping all the service facilities in the territory according to tab.1, isochrones were calculated in a GIS environment from the individual services of each category, considering the three means of transport on the respective road graphs. Next, the isochrones obtained were used to sample the census tracts by assigning a value from 0 to 3 based on their position relative to the time threshold. Consequently, tracts characterized by high accessibility (less than 5 min away from the service) were given a value of 3, going down to 1 for low accessibility sections located between 10 and 15 min from the service. The most distant census tracts are given a value of 0 since trips originating from these localities require relatively long and costly displacements. The same value (0) has also been attributed to all those census tracts that are part of a municipality in which the considered service is absent and, therefore, not available in proximity to the local inhabitants. In this way, the index simultaneously considers both the ease of access to the services available locally (accessibility dimension) and its availability in proximity (availability dimension).

The calculation was performed for all services and all municipalities, populating the reference table (Table 4.2).

The results were then summed up horizontally to identify a synthetic value associated with each census tract that expresses the overall level of accessibility to all the available services across the three modes of transport. All values obtained, which are positively influenced by the presence and good accessibility through various transport

Table 4.2 Reference table for the calculation of the accessibility by proximity index, where S = Service v = accessibility value from 0 (out of the 15 min area from service) to 3 (5 min to service, high accessibility)

Census tract	Ser. 1 (walk)	Ser.1 (car)	Ser.1 (pt)	Ser.n (w, c, pt)	Census tract synthetic result
A	vS1w in A	vS1c in A	vS1pt in A	vSnw in A	API in A = vS1w + vS1c + vS1pt + vSn...
B	vS1w in B	vS1c in B	vS1pt in B	vSnw in B	API in B = vS1w + vS1c + vS1pt + vSn...
X	vS1w in X	vS1c in X	vS1pt in X	vSnw in X	API in X = vS1w + vS1c + vS1pt + vSn...

options to the included services, were normalized and ranked in three ranges (high, medium, and low accessibility) to classify each census tract based on its accessibility by proximity performance.

Finally, the census tract level accessibility index was weighted by applying a coefficient expressing the percentage of the total municipality population that resides within the individual census tract. Subsequently, the weighted values obtained for each census tract part of the same municipality were summed to obtain a final aggregate index at the municipal level comparable with the analysis of daily im-mobility differentials presented in Chap. 3.

The calculation of the index of accessibility by proximity, both at the sub-municipal and municipal level, has returned results that confirm some already discussed territorial dynamics. More in detail, the calculation confirms the significant difference, in terms of availability and reachability of the basic services under consideration, between the municipalities of the upper mountain valleys and those located on the plains. Irrespective of the means of transport in question, the comparison between the maps of Fig. 4.2, shows that in mountain and hill municipalities service facilities are usually spatially concentrated, generating imbalances between the center and the periphery, with the most remote hamlets affected by generally low values of the accessibility index, above all because of the complex orography that makes car travel to the municipality's center complex and lengthy.

In the same mountain areas, the index at the sub-municipal scale shows how the forms of active mobility being considered (walking) can guarantee a good level of accessibility only for the inhabitants of the main urban cores and the hamlets located in their immediate vicinity. This limitation is mainly the consequence of the combined effects of the concentration of services in the central settlements and the low population density out of these centralities. These phenomena, as seen, are particularly accentuated in the Apennine municipalities, as demonstrated by the high number of dispersed hamlets with low or no pedestrian access to local services. However, this condition also affects the peri-urban areas of many municipalities of the hills and plains where the settlements outside the central cores are organized according to a reticular, denser urban structure with a higher population and settlement densities

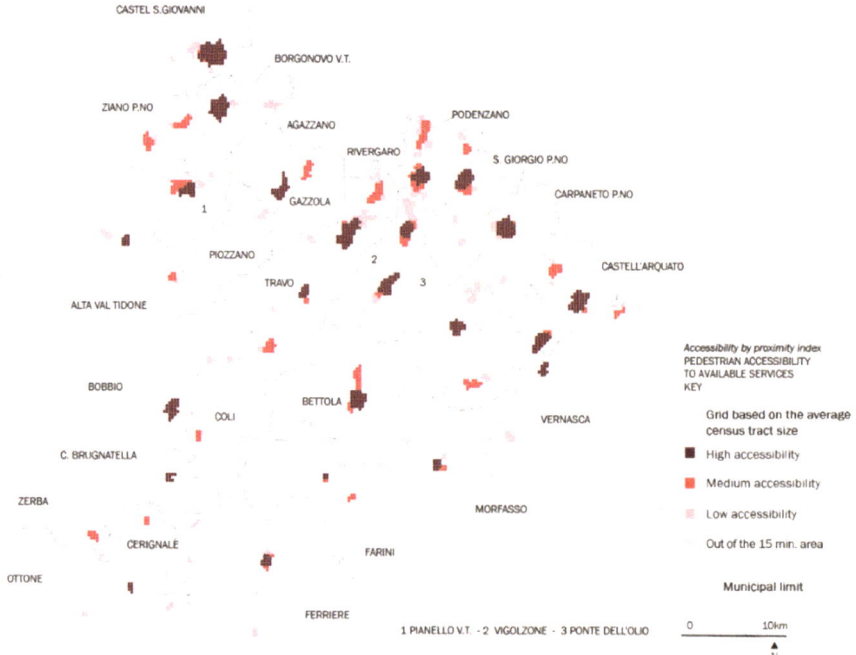

Fig. 4.2 Results of the pedestrian, car and public transport accessibility by proximity index

than in mountain areas, but with clear characteristics of dispersion and a pronounced propensity for residential mono functionality.

Outside the central cores of the municipalities throughout the whole study area, access by public transport is also strongly limited by the structure of routes and low operating frequencies. The analysis confirms the role that public transport can play in facilitating long-distance connections to Piacenza along the major roads of the valley floor, but also displays how ineffective the system can be in guaranteeing accessibility to local services moving from the peripheral hamlets of the municipalities towards the central areas of concentration of activities and services. The case of Bobbio, for example, shows how the levels of accessibility by public transport are generally good when displacing within the central core or from the hamlets along the valley floor roads towards the main center. Vice versa, moving locally from the most remote mountain hamlets towards the urban center of Bobbio becomes very complex due to the absence of transport services or their meagre daily frequencies. In the Nure valley, the DRT service solves the first problem afflicting the peripheral centers of the municipalities of the Trebbia valley, namely the lack of public transport connectivity, since this transport scheme is designed to serve all the inhabited areas of the local municipalities, whose inhabitants can rely on a capillary service—albeit infrequent.

On the contrary, the hill and plain municipalities present better levels of accessibility by public transport thanks to the good frequency of the services and the higher

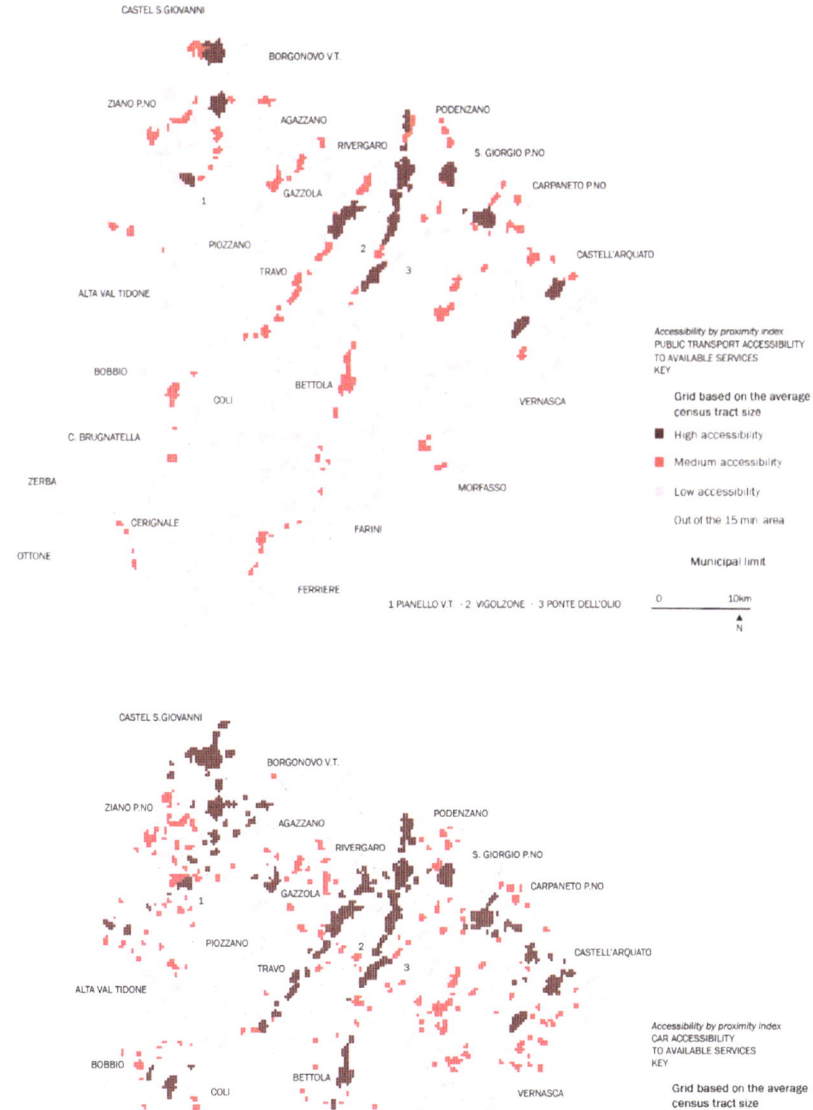

Fig. 4.2 (continued)

number of inhabitants living in the main centers where a good variety and quantity of services are available, often higher than in mountain municipalities. However, even in these realities, the public transport system is not able to reach all the inhabited areas in municipalities that, as mentioned before, are structurally characterized by a reticular network of settlements, thus making it impractical, even in these larger centers, to travel within the municipality by bus.

The analysis clearly shows that good levels of accessibility to services are guaranteed only to those who reside in the main centers and, more generally, to those who can travel by car. The availability of this means of transport is, in many cases, indispensable to guarantee activity participation to the majority of the population settled in the territory, both to move from one center to another and to reach the services available within one's municipality with relative ease. Ownership or availability of a car as a passenger, therefore, seems to be a requisite for guaranteeing access to services in a context in which the very notion of physical proximity traditionally assumed in planning cannot but be questioned by the orographic complexity of the territory, the geographical distances and the low settlement and population density. At the same time, these geographical and land-use-related factors make some services hard to reach, potentially inaccessible for those with limited motility and, at the same time, difficult to be provided in the whole territory according to a logic of physical proximity. As we have seen, these phenomena intensify in the Apennine municipalities where, net of residents living in single isolated houses outside inhabited nuclei and settlements, the majority of the population lives in areas of low or very low accessibility, as shown by Fig. 4.3 and the map depicting the results of the aggregation of the sub-municipal index at the municipal scale (Fig. 4.4).

4.4 Insights from the Accessibility Analysis and Its Integration with Immobility Measurement

The results obtained from applying the index are somewhat anticipated and heavily dependent on the choices regarding the areas of investigation, the services considered and their relevance, and the time thresholds used, which, in this experimentation, were arbitrarily selected for research purposes. However, the architecture of the proposed accessibility by proximity index allows, from a policy-oriented perspective, a comparison at the territorial level among different census tracts and municipalities. This enables the investigation of the combined effects of accessibility to services and their availability at the local scale. Moreover, the disaggregation of the results can, in turn, be very relevant for the orientation of sectoral policies: the flexibility of the method allows assessing accessibility to a selected set of activities (e.g., schools or healthcare facilities), both within the municipality and to extend the calculation to larger scales, highlighting more disadvantaged contexts that could benefit from specific improvements in direct or indirect accessibility to the selected sets of services. At the same time, excluding cars and focusing only on active and public

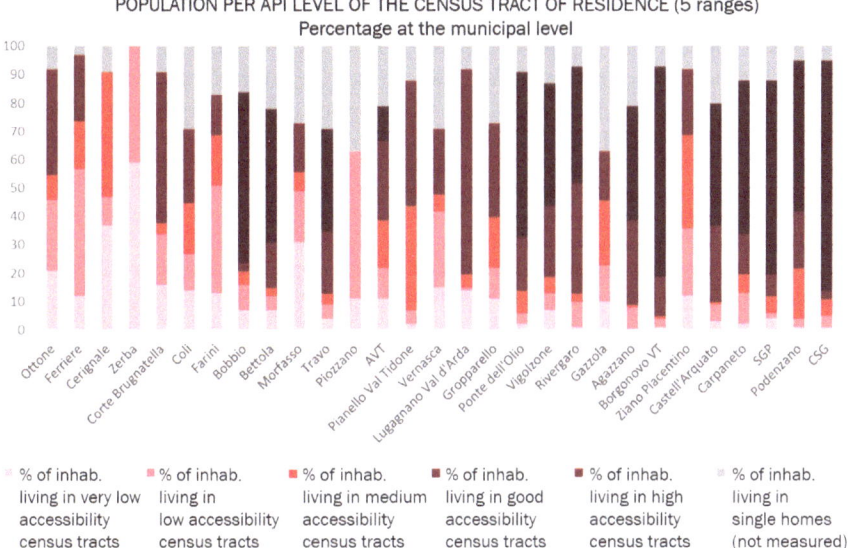

Fig. 4.3 Percentage of the population of the census tract by the level of the index (1 very low—5 very high). Municipalities are distributed on the x axis according to a geographical criterion moving from the municipalities on the south of the study area (left side, high valleys villages) to the ones on the north (right side)

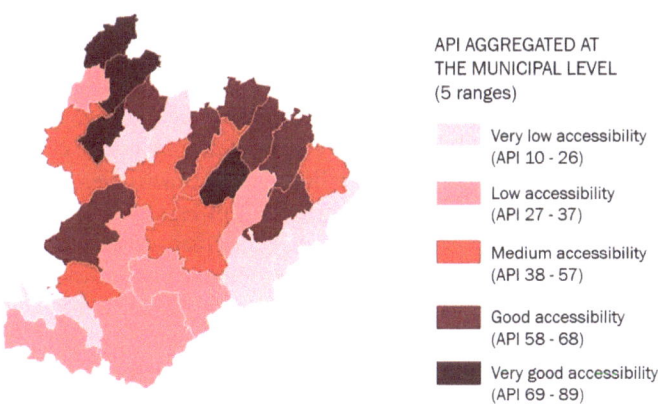

Fig. 4.4 Accessibility by proximity index aggregated at the municipal scale (weighted considering the census tract population)

transport-based modes in the calculation allow detecting all those contexts where carless inhabitants may see their possibilities of activity participation significantly limited.

From a policy-oriented perspective, the index not only allows for the analysis of territorial inequalities in accessing opportunities but also enables the simulation of the impacts of specific interventions. These interventions may concern the improvement of physical connections (e.g., new transport and mobility solutions) or the provision of specific activities and services in places where they are not currently available. Thus, the proposed approach may represent a first step towards developing an evaluative tool that can be easily adapted to different settlement and social contexts, effectively complementing other analytical tools employed for land use, transport, and welfare-related policy design and evaluation. However, the method does have some limitations that are worth discussing.

The first one concerns the definition of the geographical limits of the analysis, which are based on the administrative boundaries within which the calculation of accessibility has been confined. Although this approach can be justified is seen as a proxy for introducing the notion of physical proximity to the inhabitant in a low density context, it is certainly not able to simulate the more varied needs and necessities to which it is possible to respond by crossing one's municipality boundaries. This reasoning introduces a series of consequences. First, the index could underestimate the levels of accessibility recorded in some peripheral hamlets closer to the services provided in other locations outside the municipality. Nevertheless, since the centers of the Apennines are often lacking services and the distances between hamlets and municipalities are generally high regardless of the form and structure of administrative limits, it is possible to suppose that the boundary effect could be more intense in the municipalities of the plains and foothills, where the absence of physical barriers between settlements can make the choice of the services to be reached less constrained encouraging residents to move outside their municipality.

It is also important to note that the availability of a service facility within the municipality—hence the relative physical proximity to the service for at least a part of the population—does not imply that residents will choose or use the same facility to meet their needs. This choice could depend on the fact that some residents may not need that specific type of service, or their interest may not be compatible with the characteristics of the available service, limiting its attractiveness or usability over time. If certain activities offer standard performances regardless of location (think, for example, of pharmacies, post offices, banks), for other services other factors may come into play, such as the perceived quality or other symbolic differences that can make the closest facility less attractive and consequently redirect the choice towards other options (Næss 2006). At the same time, it is possible that performing other activities that require forms of travel may lead people to use services located outside the sphere of the proximity of their home but easily reachable, for example, from their workplaces or along the route taken during commuting, inserting these kinds of activity participation within more complex chains of daily mobility (Haugen et al. 2012).

Related to this aspect is also the fact that the list of activities and services considered in the proposed calculation represents only an approximation of the possible desires and needs of the different socioeconomic profiles of the local inhabitants, which are indeed an expression of a varied amount of opportunity types (van der Veen 2020) influenced by the structure of land use and transport, but also by individual and collective preferences. As already discussed, this last aspect is reflected not only in the definition of the activities and services to be considered, but also in the selection of the multiple factors related to the characteristics and perceptions of the quality of the paths from different users with which to characterize the simulation graphs and calculation of the index. Gathering information through qualitative methodologies about the experiences and needs of different groups of inhabitants and the ways such needs are or can be fulfilled could be a way to better contextualize—both from a spatial and social point of view—the analysis of accessibility and re-evaluate the extent and possible impacts of emerging forms of low mobility.

Another important consideration concerns the determination of the time thresholds to evaluate the levels of accessibility and of what can be considered as a condition of proximity in a low-density mountain area. In this work, thresholds of maximum 15 min were used in the awareness that this may be a questionable choice and that more extended thresholds could be used in peri urban and rural territories (e.g., the aforementioned 30-min territory). However, such thresholds should be considered as not rigid but flexibly *tailored to individual cities based on both their morphology and specific needs and characteristics* (Moreno et al. 2021, p. 106). The thresholds are, therefore, somewhat arbitrary parameters—which should be ideally verified through a sensitivity analysis—intended to "*underline that proximity-based planning is key in sustaining quality of life and in providing for the basic urban functions and that this enables works that support the spatiotemporal dimension of cities to both understand and enhance the quality of life of urban dwellers*" (ibidem). The calculation of accessibility proposed here, however, shows that Moreno's suggestive idea seems to be relatively easily applicable (at least on an analytical level) to compact urban realms but radically to be reconsidered when the basic condition of population density is lacking, as in the case of the Apennine territory of Piacenza. The critical aspect does not concern so much the space–time thresholds considered—arbitrary parameters that can be quickly modified and re-adapted—or the services included in the calculation of accessibility, but rather the actual possibility of creating more sustainable forms of accessibility by proximity and limiting the need for car-based travel and high mobility when, as demonstrated by the analysis, many inhabitants cannot but depend on cars even for the simplest and most basic daily needs. This condition concerns the territory of Piacenza and, more generally, significant portions of the country wherein, within 15, 30 or more minutes, it is (not always) possible to reach the first available service only through car-based travel that is costly for the individuals and the environment. Although the number of urban dwellers is steadily growing worldwide, a substantial portion of the global and the Italian populations still lives outside of dense urban centers: just as, following the effects of the Covid-19 pandemic, the urgency of rethinking urban life through new models and visions is perceived, so it becomes crucial to ask whether and how to promote forms of accessibility and relative immobility by proximity even in lower density environments.

Finally, a complementary use of these results may involve the comparison between the outcomes of the accessibility index at the municipal scale with those of the immobility index proposed in Chap. 3 regarding mobility differentials recorded through the variation of human presences in different municipalities. This illustrative comparison operationalizes the framework presented in Chap. 2 relating the levels of accessibility by proximity with the recorded propensity for immobility at the municipal level. Such an attempt may have analytical utility since, differently to a measure based solely on the evaluation of the levels of accessibility to basic opportunities and services, it allows for the assessment of whether a potential (accessibility) is also reflected in the intensity of an act (realized immobility), thus identifying those territorial contexts where this reflection (or its absence) can be a sign of inequality. Hence, it can be employed as a territorial scale reference to identify areas affected by dynamics that may require further investigation at a local level. This detailed investigation would still be necessary because the comparison is based on aggregated quantification that does not allow for highlighting differences and variations in the ways of being alternately mobile or immobile (the relative dimension of mobility and immobility, as discussed in Chap. 2) and how these conditions are reflected in the individual's daily life and accessibility needs, even in ways that are not necessarily linear and clear. Also, by capturing a cross-section defined over time, the method relativizes mobility differentials at the territorial level but interprets forms of relatively low mobility expressed by the total population of a single municipality as absolute. This simplification is somewhat unavoidable when performing spatial analysis on complex behaviors with aggregate quantitative data, but it does not allow exploring in a more nuanced and attentive way the variety of forms that immobility can express (Fig. 4.5).

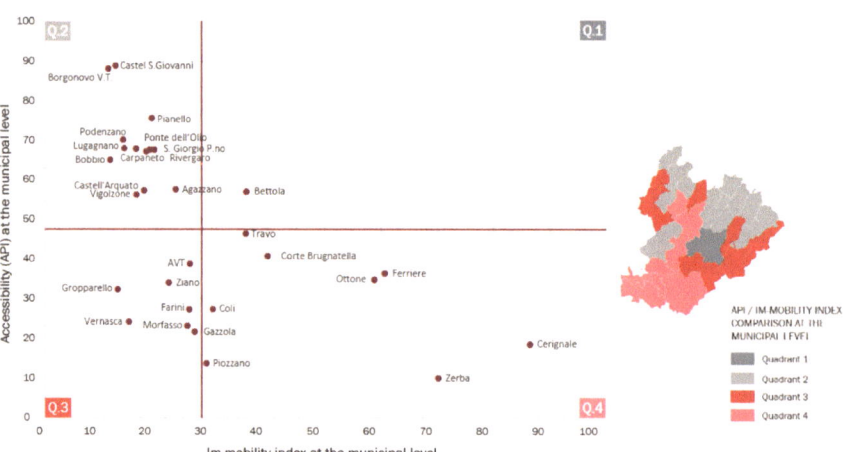

Fig. 4.5 Territorial comparison of the two indexes. A lower immobility index (x axis) describes a lower propensity for mobility (relative immobility) registered in the municipality

4.5 Conclusions

Measuring accessibility by proximity is fundamental for investigating the spatial and social conditions that can create the groundwork for forms of reversible immobility and qualifying other circumstances where the absence of nearby activities can create conditions for constrained mobility and immobility. This knowledge is highly relevant for developing specific integrated transport and land use policies that can address these conditions, if problematic, through targeted interventions which must be based on a solid analytical assessment approach. However, the theoretical density of the concept makes it challenging and complex to construct effective accessibility by proximity metrics accounting for both the spatial and individual conditions enabling reversible immobility in relation to the variety of possible accessibility needs and the means available for meeting them in terms of transport and active transport modes. These already challenging issues may become even more complex when analysing territorial contexts that, by their nature, require a reinterpretation of the principle of proximity, such as areas characterized by low-density population and activities. In these contexts, similar to the case study presented in the chapter, new accessibility metrics and measurement tools need to be experimented with to extend a planning principle, namely the promotion of accessibility by proximity, where proximity between inhabitants and services may not be guaranteed due to economies of scale, dispersed settlements, and shrinking numbers of service users.

The experimentation conducted in the Piacenza Apennines has shown that accessibility by proximity in the territory is particularly low, especially in areas where it would be most important to ensure it, namely the more remote, aging and economically less developed areas. Conversely, only a predominantly car-based and relatively high distance mobility allows ensuring a minimum level of accessibility to valley services.

This awareness leads to questioning whether reversible immobility can effectively take place in this territory. It also raises questions about the perceptions, difficulties, and needs of both immobile inhabitants who might suffer from constrained immobility and mobile inhabitants who might perceive their mobility as constrained by the specific circumstances of low proximity to services and opportunities. These aspects related to the dimension of individual experience are difficult to interrogate through a quantitative approach to accessibility measurement as proposed in this chapter. This consideration highlights the utility of complementing this investigation with a deeper qualitative study that focuses on the experiences of individuals and the communities they represent, as proposed in the following chapter.

References

Arellana J, Saltarin M, Larranaga AM, Gonzalez VB, Henao CA (2020) Developing an urban bikeability index for different types of cyclists as a tool to prioritise bicycle infrastructure investments. Transp Res Part A-Policy Pract 139:310–334. https://doi.org/10.1016/j.tra.2020. 07.010

Benenson I, Martens K, Rofé Y (2010) Measuring the gap between car and transit accessibility: estimating access using a high-resolution transit network geographic information system. Transp Res Rec 2144:28–35. https://doi.org/10.3141/2144-04

Benenson I, Ben-Elia E, Rofe Y, Rosental A (2017) Estimation of urban transport accessibility at the spatial resolution of an individual traveler. Springer Geography 383–404. https://doi.org/10. 1007/978-3-319-40902-3_21

Beria P (2020) Quale mobilità durante e dopo il COVID19 nei territori fragili? Blog del Dipartimento di Eccellenza sulle fragilità territoriali. Available at: https://www.eccellenza.dastu.polimi.it/wp-content/uploads/2020/05/Beria-2020-Mobilità-covidterritori-fragili.pdf

Boeing G (2017) OSMnx: new methods for acquiring, constructing, analyzing, and visualizing complex street networks. Comput Environ Urban Syst 65:126–139. https://doi.org/10.1016/j. compenvurbsys.2017.05.004

Boisjoly G, El-Geneidy AM (2017a) How to get there? A critical assessment of accessibility objectives and indicators in metropolitan transportation plans. Transport Policy 55(October 2016):38–50. https://doi.org/10.1016/j.tranpol.2016.12.011

Boisjoly G, El-Geneidy AM (2017b) The insider: a planners' perspective on accessibility. J Transp Geogr 64:33–43. https://doi.org/10.1016/j.jtrangeo.2017.08.006

Buck C, Pohlabeln H, Huybrechts I, De Bourdeaudhuij I, Pitsiladis Y, Reisch L, Consortium I (2011) Development and application of a moveability index to quantify possibilities for physical activity in the built environment of children. Health Place 17:1191–1201. https://doi.org/10.1016/j.hea lthplace.2011.08.011

Büttner B, Seisenberger S, Baquesro Larriva MT, Rivas De Gante AG, Haxhija S, Ramirez A, McCormick B (2022) Urban mobility next 9 ±15-minute city: human-centred planning in action mobility for more liveable urban spaces. Available at: https://www.eiturbanmobility.eu/wp-con tent/uploads/2022/11/EIT-UrbanMobilityNext9_15-min-City_144dpi.pdf. 2022

Campisi T, Ignaccolo M, Inturri G, Tesoriere G, Torrisi V (2021) Evaluation of walkability and mobility requirements of visually impaired people in urban spaces. Res Transp Bus Manag 40:100592. https://doi.org/10.1016/j.rtbm.2020.100592

Clifton KJ, Livi Smith AD, Rodriguez D (2007) The development and testing of an audit for the pedestrian environment 80(1–2):0–110. https://doi.org/10.1016/j.landurbplan.2006.06.008

Currie G, Delbosc A (2011) Mobility versus affordability as motivations for car-ownership choice in urban fringe, low-income Australia. In: Lucas K, Blumenberg E, Weinberger R (eds) Auto motives. understanding car use behaviours, pp 193–208. Emerald, Bingley

Erath A, Van Eggermond MA, Ordonez S, Axhausen KW (2017) Introducing the pedestrian accessibility tool: Walkability analysis for a geographic information system. Transp Res Rec 2661(1):51–61. https://doi.org/10.3141/2661-06

Ewing R, Handy S (2009) Measuring the unmeasurable: urban design qualities related to walkability. J Urban des 14(1):65–84. https://doi.org/10.1080/13574800802451155

Farber S, Morang MZ, Widener MJ (2014) Temporal variability in transit-based accessibility to supermarkets. Appl Geogr 53:149–159. https://doi.org/10.1016/j.apgeog.2014.06.012

Farrington JH (2007) The new narrative of accessibility: its potential contribution to discourses in (transport) geography. J Transp Geogr 15(5):319–330. https://doi.org/10.1016/j.jtrangeo.2006. 11.007

Farrington J, Farrington C (2005) Rural accessibility, social inclusion and social justice: towards conceptualization. J Transp Geogr 13:1–12

Frank LD, Sallis JF, Saelens BE, Leary L, Cain K, Conway TL, Hess PM (2010) The development of a walkability index: application to the neighborhood quality of life study. Br J Sports Med 44(13):924–933. https://doi.org/10.1136/bjsm.2009.058701

Gaglione F, Gargiulo C, Zucaro F (2022) Where can the elderly walk? A spatial multicriteria method to increase urban pedestrian accessibility. Cities 127:103724. https://doi.org/10.1016/j.cities.2022.103724

García-Albertos P, Picornell M, Salas-Olmedo MH, Gutiérrez J (2019) Exploring the potential of mobile phone records and online route planners for dynamic accessibility analysis. Transp Res Part A: Policy Pract 125:294–307. https://doi.org/10.1016/j.tra.2018.02.008

Geurs KT, van Wee B (2004) Accessibility evaluation of land-use and transport strategies: review and research directions. J Transp Geogr 12(2):127–140. https://doi.org/10.1016/j.jtrangeo.2003.10.005

Guzman LA, Oviedo D (2018) Accessibility, affordability and equity: assessing 'pro-poor' public transport subsidies in Bogotá. Transp Policy 68(April):37–51. https://doi.org/10.1016/j.tranpol.2018.04.012

Guzman LA., Oviedo D, Rivera C (2017) Assessing equity in transport accessibility to work and study: The Bogotá region. J Transp Geogr 58:236–246. https://doi.org/10.1016/j.jtrangeo.2016.12.016

Handy S (2020) Is accessibility an idea whose time has finally come? Transp Res Part D: Transp Environ 83(April):102319. https://doi.org/10.1016/j.trd.2020.102319

Hansen WG (1959) How accessibility shapes land-use. J Am Inst Plann 25(2):73–76

Haugen K, Holm E, Strömgren M, Vilhelmson B, Westin K (2012) Proximity, accessibility and choice: a matter of taste or condition? Pap Reg Sci 91:65–84. https://doi.org/10.1111/j.1435-5957.2011.00374.x

Iacono M, Krizek KJ, El-Geneidy A (2010) Measuring non-motorized accessibility: issues, alternatives, and execution 18(1):0–140. https://doi.org/10.1016/j.jtrangeo.2009.02.002

Järv O, Ahas R, Witlox F (2014) Understanding monthly variability in human activity spaces: a twelve-month study using mobile phone call detail records. Transp Res Part C: Emerg Technol 38:122–135. https://doi.org/10.1016/j.trc.2013.11.003

Järv O, Tenkanen H, Salonen M, Ahas R, Toivonen T (2018) Dynamic cities: location-based accessibility modelling as a function of time. Appl Geogr 95(May):101–110. https://doi.org/10.1016/j.apgeog.2018.04.009

Jones P, Lucas K (2012) The social consequences of transport decision-making: clarifying concepts, synthesising knowledge and assessing implications. J Transp Geogr 21:4–16. https://doi.org/10.1016/j.jtrangeo.2012.01.012

Kamruzzaman M, Yigitcanlar T, Yang J, Mohamed MA (2016) Measures of transport-related social exclusion: a critical review of the literature. Sustainability (Switzerland) 8(7):6–11. https://doi.org/10.3390/su8070696

Kenyon S, Lyons G, Rafferty J (2002) Transport and social exclusion: investigating the possibility of promoting inclusion through virtual mobility. J Transp Geogr 10(3):207–219

Lanza G, Pucci P, Carboni L (2023) Measuring accessibility by proximity for an inclusive city. Cities 143:104581. https://doi.org/10.1016/j.cities.2023.104581

Lanza G, Pucci P (2022) Distributing, DesynchroniSing, DigitaliSing: towards a new mobile urbanity in the COVID-19 era. In: Balducci S, Armondi S, Bovo M, Galimberti B (eds) Cities learning from a pandemic: towards preparedness. Routledge

Lanza G, Pucci P, Carboni L, Vendemmia B (2022) Impacts of the Covid-19 pandemic in inner areas. Tema. J Land Use Mobil Environ 73–89. https://doi.org/10.6092/1970-9870/8915

Levine J, Grengs J, Merlin LA (2019) From mobility to accessibility. Transform urban transportation and land use planning. Cornell University press, Ithaca (NY)

Li Q, Zhang T, Wang H, Zeng Z (2011) Dynamic accessibility mapping using floating car data: a network-constrained density estimation approach. J Transp Geogr 19(3):379–393. https://doi.org/10.1016/j.jtrangeo.2010.07.003

Lowry MB, Callister D, Gresham M, Moore B (2012) Assessment of communitywide bikeability with bicycle level of service. Transp Res Rec 2012(2314):41–48

Lucas K (2012) Transport and social exclusion: where are we now? Transp Policy 20:105–113

Lucas K, van Wee B, Maat K (2016) A method to evaluate equitable accessibility: combining ethical theories and accessibility based approaches. Transportation 43:473–490

Martens K (2017) Transport justice. Designing fair transportation systems. Routledge, Abingdon

Martens K, Di Ciommo F, Papanikolaou A (2014) Incorporating equity into transport planning: utility, priority and sufficiency approaches. In: XVIII Congreso Panamericano de Ingeniería de Tránsito, Transporte y Logística, Santander, 11–13th June

Martens K (2016) Why accessibility measurement is not merely an option but an absolute necessity. In: Punto N, Hull A (eds) Accessibility tools and their applications. Routledge

Mattioli G (2014) Where sustainable transport and social exclusion meet: households without cars and car dependence in Great Britain. J Environ Planning Policy Manag 16(3):379–400. https://doi.org/10.1080/1523908X.2013.858592

Mondschein A, Taylor BD (2017) Is traffic congestion overrated? Examining the highly variable effects of congestion on travel and accessibility. J Transp Geogr 64(June):65–76. https://doi.org/10.1016/j.jtrangeo.2017.08.007

Moreno C, Allam Z, Chabaud D, Gall C, Pratlong F (2021) Introducing the "15-minute city": sustainability, resilience and place identity in future post-pandemic cities. Smart Cities 4(1):93–111. https://doi.org/10.3390/smartcities4010006

Næss P (2006) Accessibility, activity participation and location of activities: exploring the links between residential location and travel behaviour. Urban Studies 43:627–652

Niedzielski MA, Eric Boschmann E (2014) Travel time and distance as relative accessibility in the journey to work. Ann Assoc Am Geogr 104(6):1156–1182. https://doi.org/10.1080/00045608.2014.958398

Pajares E, Büttner B, Jehle U, Nichols A, Wulfhorst G (2021) Accessibility by proximity: addressing the lack of interactive accessibility instruments for active mobility. J Transp Geogr 93(C). https://doi.org/10.5198/jtlu.2015.585

Papa E, te Brömmelstroet M, Silva C, Hull A (2016) Accessibility instruments for planning practice: a review of European experiences. J Transp Land Use 9(3):57–75. https://doi.org/10.5198/jtlu.2015.585

Pinna F, Murrau R (2018) Age factor and pedestrian speed on sidewalks. Sustainability (Switzerland) 10(11). https://doi.org/10.3390/su10114084

Poorthuis A, Zook M (2023) Moving the 15-minute city beyond the urban core: the role of accessibility and public transport in the Netherlands. J Transp Geogr 110:103629. https://doi.org/10.1016/j.jtrangeo.2023.103629

Portegijs E, Keskinen KE, Tsai L, Rantanen T, Rantakokko M (2017) Physical limitations, walkability, perceived environmental facilitators and physical activity of older adults in Finland. Int J Environ Res Public Health 14(3):333. https://doi.org/10.3390/ijerph14030333

Preston J, Rajé F (2007) Accessibility, mobility and transport-related social exclusion. J Transp Geogr 15(3):151–160. https://doi.org/10.1016/j.jtrangeo.2006.05.002

Pucci P, Vecchio G (2019) Enabling mobilities. Springer, PoliMi Springer Brief

Pucci P, Carboni L, Lanza G (2022a) Accessibilità di prossimità in un territorio montano. ASUR, Archivio di Studi Urbani e Regionali, 135–138

Pucci P, Carboni L, Lanza G (2022b) Accessibilità di prossimità per una città più equa. Sperimentazione in un quartiere di Milano. TERRITORIO (99):40–52. https://doi.org/10.3280/tr2021-099006

Shergold I, Parkhurst G, Musselwhite C (2012) Rural car dependence: an emerging barrier to community activity for older people. Transp Plan Technol 35(1):69–85. https://doi.org/10.1080/03081060.2012.635417

Silva C, Bertolini L, te Brommelstroet M, Milakis D, Papa E (2017) Accessibility instruments in planning practice: Bridging the implementation gap. Transp Policy 53:135–145. https://doi.org/10.1016/j.tranpol.2016.09.006

SNAI Inner area report (2018–2020)—Appennino smart. Adattamento intelligente per cambiare gli schemi d'azione e superare le criticità con nuove idee. Available at https://www.agenziaco esione.gov.it/wpcontent/uploads/2020/07/Definitivo_Strategia_AI_Appennino_PC_PR_dic_ 18.pdf. Accessed 11 Apr 2022

Southworth M (2005) Designing the walkable city. J Urban Plan Dev 131(4):246–257. https://doi. org/10.1061/(ASCE)0733-9488(2005)131:4(246)

Talavera-Garcia R, Soria-Lara JA (2015) Q-PLOS, developing an alternative walking index. A method based on urban design quality. Cities 45:7–17. https://doi.org/10.1016/j.cities.2015. 03.003

te Brömmelstroet M (2010) Equip the warrior instead of manning the equipment: land use and transport planning support in the Netherlands. J Transp Land Use 3(1):25–41

van der Veen AS, Annema JA, Martens K, van Arem B, Correia GHA (2020) Operationalizing an indicator of sufficient accessibility—a case study for the city of Rotterdam. Case Stud Transp Policy 8(4):1360–1370. https://doi.org/10.1016/j.cstp.2020.09.007

van Wee B, Geurs K, Chorus C (2013) Information, communication, travel behavior and accessibility. J Transp Land Use 6(3):1–16. https://doi.org/10.5198/jtlu.v6i3.282

Vecchio G (2019) Accessibility: enablement by access to valued opportunities. In: Pucci P, Vecchio G (eds) Enabling mobilities, planning tools for people and their mobility. Springer, Cham

Vitale Brovarone E, Cotella G (2020) Improving rural accessibility: a multilayer approach. Sustainability (Switzerland) 12(7). https://doi.org/10.3390/su12072876

Witten K, Pearce J, Day P (2011) Neighbourhood destination accessibility index: a GIS tool for measuring infrastructure support for neighbourhood physical activity. Environ Plan A 43(1):205–223. https://doi.org/10.1068/a43219

Chapter 5
Analysing Individual Experiences of Immobility

Abstract This chapter examines immobilities as experienced at both individual and collective levels, along with the factors influencing related choices or constraints. Utilizing the technique of microstories within a specific case study, it investigates the individual causes and effects of immobility in relation to local accessibility conditions. This approach deepens the understanding of immobility presented in the theoretical framework of Chap. 2, demonstrating how mobility and immobility can be perceived as reversible, chosen, or constrained, depending on the interplay between contextual factors and personal experiences. Additionally, the chapter identifies a set of so-defined reversibility factors—shaped by public policies, community collaboration, and individual perceptions of access to immaterial goods—that may facilitate a transition from constrained conditions to chosen, reversible mobility and immobility. Consequently, the concepts of mobility and immobility enablement are introduced to highlight the potential roles of public policies and community action in activating these reversibility factors.

5.1 Introduction

In the previous chapters, different tools of a purely quantitative nature were described and applied to evaluate how various realized levels of immobility and travel may or may not be associated with forms of potential disadvantage if traced back to the accessibility by proximity conditions that contribute to defining part of the extent of opportunities and capabilities available to an individual. Referring, to Sen's (2005) theory of the capability approach and the role of capabilities in shaping a person's mobility behavior and social participation, the existence of differentials in the daily mobility practices of individuals generating forms of relative immobility may be due to a reduced ability to turn one's resources and capabilities (possibility to access opportunities) into a specific functioning (realized mobility) (Vecchio and Martens 2021). However, as illustrated in the previous chapters, it is argued that such an outcome could also be the expression of different modes of participation enabled by the influence, on an individual's capabilities, of the physical proximity and accessibility to needed and desired activities and opportunities. Thus, in this perspective, relative immobility can be determined not only by the inability of an individual to translate their resources into a capability and then functioning due to the action of different constraints, but also when the achieved access to a specific opportunity in proximity can push towards choices and behaviors consisting of limited travel without compromising activity participation.

However, it is essential to emphasize that the approach proposed in previous phases of the research does not intend to identify specific causal links between conditions of accessibility influencing a person's capabilities and realized immobility. Referring to the points raised by Nussbaum (2003), rather than a single type of capability, it would be appropriate to speak of plural capabilities in framing the diversified ways in which individuals translate personal and contextual resources into actual different functionings. These plural modes of translation inevitably reflect on the sphere of mobility, giving rise to very different experiences, behaviors, and practices between individuals even in the presence of similar contextual conditions related, for example, to the availability of transportation systems and activities, including those accessible in physical proximity (Cao and Hickman 2019). Thus, the dimension of individual experience becomes a central element in understanding the factors that can shape a person's needs and potential for access, as well as in how this potential translates or not (whether by choice or by the effect of constraints) into functioning. This awareness leads to orient research towards an approach focused on the individual sphere, to explore the manifold expressions of immobility practices experienced by different people in relation to their social and spatial contexts of belongings.

Four fundamental issues regarding the translation of the concept of immobility into urban planning have been identified in Chap. 2. Here, the focus is specifically set on the third point regarding the analysis of the individual experience of immobility. This last aspect is explored discussing the role of narrative inquiry techniques and, more specifically, microstories collection, as a powerful research tool to investigate the effects of individual experiences of immobility and the reasons behind related choices or constraints. This qualitative research methodology and the insights resulting from its implementation are specifically discussed in their ability to integrating the results obtained from the quantitative assessments illustrated in the previous chapters. As discussed, purely quantitative methods can highlight general trends at the aggregated social and spatial level but not shed light on the complexity of the forms and implications of mobility differentials and the role of micro-mobility in ensuring the social participation of the individual. Conversely, qualitative techniques can contribute, in a concrete case study, to a broader and more nuanced understanding of the social and spatial implications of immobility. In this perspective, the chapter presents the results of qualitative on-site fieldwork conducted in the municipalities of the Apennine area of the province of Piacenza, focusing on the relevant insights provided by this direct research experience. After introducing the tool of microstories and its methodological application in the case study (Sect. 5.2), several stories of daily immobility from the inhabitants of the most remote and inaccessible areas of the territory are presented, divided into thematic profiling (Sect. 5.3). This description is followed by a summary of the different conditions that emerged and the contribution that a microstories-based approach can offer to shed light on what have been defined as reversibility factors, namely the variable conditions induced by local policies and the collaborative actions of community members in supporting mobility and accessibility, that can contribute to reversing constrained mobility and immobility (Sect. 5.4). Conclusive remarks close the chapter.

5.2   Microstories: A Tool to Explore the Individual Implications of Immobility

Microstories collection is a methodology that can be ascribed to the so-called narrative turn, which has led to the concurrent development of narrative research inquiry in qualitative research. The latter, defined by Polkinghorne (1995) as *"a subset of qualitative research designs in which stories are used to describe human action"* (1995, p. 5), has emerged within new reflections in philosophical discussions about the relationships *"between self, other, community, social, political and historical dynamics (…) questioning and challenging the positivist approach to examining the social world and understanding human experience"* (Goodson and Scherto 2011, p. 18). Microstories collection fits within these techniques by making use of individual stories told by the interviewees and conversations collected during participant observation by the researcher. When microstories are collected and subsequently subjected to editing, interpretation, and presentation, even in conjunction with other sources, they can be considered part of a "life history" approach according to the systematization proposed by Ojermark (2007) on narrative research inquiry. The utility of microstories as a research technique lies in the fact that their collection and interpretation becomes a way to *"respond - at least in part - to the need of describing individuals' experiences, perceptions and values, going beyond aggregate approaches and enriching the understanding of how everyday mobility practices contribute to one's overall participation in social life"* (Vecchio 2020, p. 3). Focusing the analysis at the individual scale enriches and complements the quantitative measurements by showing complex phenomena that aggregate analyses often return partially. Moreover, microstories offer the opportunity to challenge the results of quantitative analysis and test them at a different scale, observing the variations and 'particular' exceptions to be analyzed and contextualized to define different profiles of immobility.

This research uses microstories primarily to explore the variety of immobility practices and experiences in particular territorial and social contexts. However, microstories are also used to discuss how the comparison between the experiences and points of view collected during the interviews could provide practical elements to rethink the same quantitative analyses previously proposed, ultimately contributing to enhance their sensitivity to the social and spatial context of study. This updating can be done, for example, by introducing information that a desk-based study cannot capture without a direct exchange with residents and stakeholders and by reconsidering some critical aspects such as the definition of the inhabitants' needs and how these are or are not satisfied through the immobility practices described by the interviewees. A qualitative approach based on public engagement can help (re)orient integrated transportation and land use planning policies aimed at promoting equity and ensuring accessibility to services and opportunities *"accounting for differential voices, knowledge, experiences, abilities, and rhythms of the actors that inhabit particular spaces and places showing how these differences are enabled or constrained by policy frameworks themselves"* (Verlinghieri and Schwanen 2020), especially in a context to which are attributed characters of social and spatial marginality. Still, the possibility that a more qualitative approach can be used under this perspective clearly depends on the construction of more structured methods of public engagement than those proposed in this experimental research.

The on-site research presented in this chapter investigates some characteristics of the immobility practices of the inhabitants and how they enable or disable forms of activity participation based on their accessibility by proximity needs. In addition to the inhabitants, some key informants and local stakeholders were interviewed to build a preliminary picture of the specific conditions affecting the livability of a territory considered marginal and collect their views and proposals about the ongoing and planned policies and actions for the revitalization of the area. Direct interaction with key informants and residents made it possible to discover a whole series of initiatives to support the most fragile population segments and (re)activate the territory both top-down and bottom-up in nature that, for their very local relevance, were previously unknown by the researcher. While the former initiatives mainly consist of micro-scale welfare or mobility policies that are generally implemented at the provincial level or by single municipalities, the latter refers to the existence of systems of mutual support and help that develop spontaneously within local communities This latter point will be broadly discussed, considering the fact that accessibility, rather than being only an individual, transport related or spatial issue, could be interpreted as a collective matter. This is the case when accessibility needs are met through local collaboration and participation enabled by relational proximity between people, (Manzini 2021) often to respond to lack of spatial and functional proximity to needed activities and opportunities. This aspect will be deepened in Sect. 5.4.

Inspired by the plurality of ethnographic techniques on mobility practices that can be traced back to the emergence of the so-called mobile methods and the narrative turn in qualitative research, different tools have been used for the investigation and microstories outlining: direct observation, face-to-face semi-structured interviews with key informants, surveys to various members of the community, and mapping.

The empirical analysis focuses on two areas of the province of Piacenza, the upper Val Nure, and Val Trebbia. The municipalities of the upper valleys were chosen because they are characterized by socioeconomic, demographic, and geographic conditions that qualify them as municipalities affected by various forms of marginality, potentially immobile, and very inaccessible (See the results from Chaps. 3 and 4). However, the survey on the variation of human presences proposed in Chap. 3 somewhat challenged this reading, indicating these municipalities as characterized by forms of constrained mobility, with high rates of daily mobility expressed by residents associated with deficient levels of physical accessibility to services and opportunities. The apparent contradiction that emerged between the two results led the researcher to deepen the analysis by disaggregating the level of observation, looking for conditions of relatively low and high mobility within the same area but expressed by different individuals, forms of interdependence between mobile and immobile subjects, different temporal scans, by the same individual, in the intermittence between moments of relative immobility and mobility.

For the Val Trebbia, the focus is on two municipalities of the high valley (Zerba and Cerignale) as those resulted as the most remote, marginal, and poorly accessible. For the Val Nure, Bettola and its many dispersed hamlets are investigated to analyze the effect of the settlement's spatial dispersion on the immobility practices of the inhabitants. As discussed in the previous chapter, the aggregate analyses at the municipal level are unable to delve into the complexity of micro-mobilities that might occur between sub-areas of the same municipality. Instead, the qualitative approach offers this opportunity.

Concerning the survey with the inhabitants, these were not previously organized but were carried out through a direct approach intercepting people in public spaces or within commercial activities (Vendemmia and Lanza 2022). From the initial interviews, it was possible to extend the sample through an incremental snowballing technique thanks to the indication, both by the interviewed inhabitants and key informants, of other people to be interviewed and involved in the research. Regarding the sample's composition (n = 14), most of the interviewed are of advanced age (over 75 y.o.) (n = 6), mainly due to the area's demographic composition. At the same time, while traveling around the context, the sample was enlarged by including other subjects of different age profiles (n = 8), which allowed deepening the knowledge of the territory and the needs of its inhabitants. In some cases, these younger people were interviewed because their daily activities and immobility practices are influenced by interdependence with other mobile or immobile people (e.g., caregivers). In other cases, because these people are part of demographically under-represented segments of the population who live and reside in these areas but may not be considered primary targets for welfare or mobility policies in such high ageing contexts. Therefore, extending the sample to a broader group represents an opportunity to explore the everyday immobility practices of different population segments within the same area of analysis and how they perform them. Together with the semi-structured interviews, other shorter conversations were conducted with inhabitants and users of the region that allowed us to extract further information on living conditions in the area (Fig. 5.1).

After collecting data on their social and economic condition, the interview focused on understanding their needs, preferred means of transport, how often they need to travel to reach the opportunities required and how far they need to go. The interview was composed of different sections: (1) general information on the subject; (2) household condition; (3) mobility competencies; (4) daily activities, including internet-based ones and mobility practices; (5) means of transport used; (6) experience during the Covid-19 pandemic lockdown. The interviews were conducted

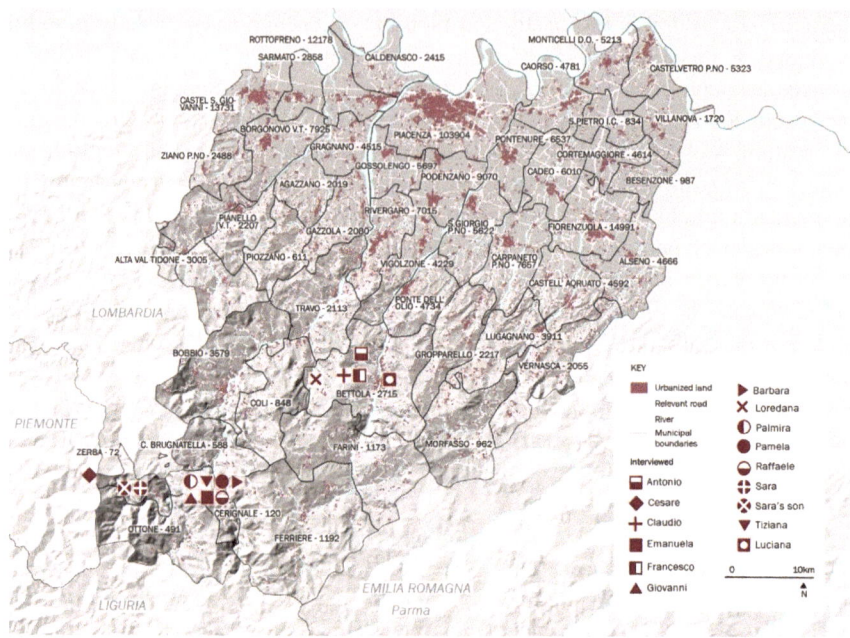

Fig. 5.1 Locations of the interviews. Respondents' names are pseudonymized for privacy concerns

face-to-face because most of the interviewees, particularly the elderly, were not even reachable over the phone or via e-mail as they had no accessibility to a landline, mobile phone, or internet connection.

Within the research method, the microstories become a promising tool, but still subject to limitations, to capture new elements of discussion and interpretation. The proposal that is made is that the direct involvement of mobility (and immobility)-related policy recipients is essential to recognize the plurality of conditions, needs, and effects related to existing forms of mobility and immobility in different territorial contexts. From this point of view, a structured involvement of local communities, of which this work represents limited experimentation, could fruitfully integrate quantitative approaches such as those proposed in this research towards a greater sensitivity to the peculiarities of each local context.

Paraphrasing Bissell (2018), the contents of this chapter, while exploring some issues from a policy-oriented perspective, do not constitute an indicative representation of wider things as much as creative encounters themselves, offering the potential for becoming attuned to the uniqueness of different situations.

5.3 Being Immobile: Practices and Experiences in the Piacenza Apennines

In this paragraph, the results of the interviews are presented, organized according to thematic lines that allow defining different profiles and conditions of immobility concerning the daily experiences of the inhabitants and territory users. Profiles of immobility are thus built based on the conceptual framework introduced in Chap. 2. Inspired by the scientific literature, the scheme envisaged four situations linked to the degree of constraint or reversibility with which forms of mobility and immobility could be experienced based on the level of accessibility to services and opportunities. Together with the four categories of constrained and reversible immobilities, another profile of so-called immobility enablers has been added, that is, all those subjects that support the immobilities of others through their daily practices, creating complex and interesting forms of interdependence that deserve a specific in-depth account.

The qualitative approach has favored overcoming a static schematization of the existing conditions, identifying specific factors of several types, defined as *reversibility factors*, which facilitate the transition from constrained forms of mobility and immobility that may limit activity participation towards forms of mobility and immobility perceived as reversible. Indeed, a transition between constrained and reversible forms due to the action of a plurality of factors foreshadows the role of planning policies in directly or indirectly influencing- at least some- of these factors, as discussed at the end of the chapter.

5.3.1 Constrained Immobile People

A constraint can be defined as something that keeps a person within particular limits due to the action of specific factors that, for various reasons, prevent or restrict the ability to access what lies beyond these limits. Immobility may thus result from the existence of constraint factors that contribute to restrict the extent of an individual's area of daily activity. Age, health, and physical autonomy are among the most relevant ones as they contribute to defining independence in one's mobility practices and behaviors.

In the small village of Cerignale, several very elderly inhabitants, including Tiziana, Emanuela, Giovanni, and Pamela, all over 75 years old, describe a minimal spatial area of daily activity.

For Giovanni and Pamela, this area coincides with their apartments. Both live in the same building, without an elevator, a few steps from the village's town hall and the food store. Due to age-related impairments, they find it challenging to move from one room to another and only go out in exceptional cases, always accompanied by others. Their limited mobility enrols them in the home care system provided by the local health authority, receiving visits from community nurses and doctors several times a week, as needed.

Emanuela's situation differs. She lives on the ground floor of an older house behind the town hall, amid the small and narrow cobblestone streets of the village's historic core. Emanuela manages to leave her house to reach the church, the local clinic set up in a room in the town hall where the general practitioner sees patients a few days a week, and the small food store. Although these activities are in close proximity and require only short trips, reaching them is increasingly difficult due to the steep terrain, uneven pavement, and a sense of fatigue, making Emanuela less confident in her ability to move and increasingly reluctant to go out. Emanuela's condition indicates a progressive narrowing of her daily area of activity and participation, which, in some ways, could be considered as chosen because it is more reassuring in response to the concerns and feelings of vulnerability induced by leaving her home.

In contrast, Tiziana continues to work despite her advanced age. She lives and works in the same building, operating accommodation and restaurant activities with seasonal helpers for a few months a year. During the annual opening periods, her subsistence needs are met through deliveries made by restaurant suppliers. During the closing periods, her skills allow her to reach the food store, located just under 100 m from her building's entrance, and the local doctor's office. Twice a year, she travels to the valley, generally to Ottone but preferably to the larger center of Bobbio, for hairdressing and purchasing clothes or personal care items. For these trips, she is always accompanied by family members or uses the social taxi system, a public on-demand transit service organized by the Municipality of Cerignale and dedicated to elderly people to facilitate their displacements to fulfill specific needs such as reaching specialized healthcare facilities, commercial activities, public offices and social hotspots that are far from the user's place of residence,

The experiences of Tiziana, Emanuela, Giovanni, and Pamela share a reliance on limited mobility, leading to a condition of relative immobility where daily activities are conducted within a very restricted space, aligned with their limited resources and capabilities. All four older adults can be considered immobile even if each of them experiences immobility differently based on the intensity of physical and perceptual constraints affecting their micro-mobility. The physical abilities of the respondents influence their displacements: especially in Giovanni and Pamela's cases, the small daily movements they perform between rooms require effort, concentration, and energy expenditure, making their apparent daily immobility an alternation of moments of stasis and potentially intense micro-mobility. Conversely, Emanuela and Tiziana have better movement capabilities, although progressively reduced for Emanuela. Tiziana uses her abilities to engage in local activities and interact with other inhabitants near the restaurant, making her relatively less immobile than Emanuela, Giovanni, and Pamela.

However, it is worth considering whether the physical constraints related to age, albeit with varying intensities, lead to a downsizing of expectations and needs, prompting a reconsideration of priorities and acceptance that the quantity, variety, and spatial distribution of activities may gradually shrink according to the logic of adaptive preferences (Vecchio 2020). This induced and chosen downsizing raises the question of whether the immobility of these individuals should be considered constrained, regardless of their territorial living contexts. On one hand, their relative immobility could be seen as the natural consequence of choices induced by perceived decreasing capabilities and possibilities. On the other, because their immobility does not seem to preclude them from accessing some limited resources that are still essential for their well-being and in which they still manage to participate, often thanks to the help of others.

In fact, older individuals' immobility could not occur without the mobilization of other essential figures who provide assistance and daily subsistence. A relevant point that emerged from the interviews relates to family caregiving, which, while allowing the most fragile family members to remain immobile, creates unique forms of immobility. Palmira, Pamela's daughter, was born and lives in Cerignale, returning after retirement following years of living elsewhere. She lives near her elderly parents' home, offering daily assistance alongside Paula, a caregiver from South America who stays in Palmira's parents' apartment. Palmira cannot drive and cannot rely on her husband, who can no longer drive due to disability. Consequently, Palmira must organize daily care and subsistence activities, leveraging the few local opportunities and limiting her mobility away from Cerignale due to the difficulty of moving without a car in the mountainous territory. The presence of a nearby food store, opened a few years ago and operating only during certain hours in winter, is essential for balancing parental care and ensuring their livelihood without a private vehicle. The availability of Cerignale's social taxi, used for infrequent trips to valley centers like Coli or Ottone, and the support from other village inhabitants who offer rides, also compensate for the inadequate public transport system. The only bus near her house runs infrequently at impractical times and takes a long, uncomfortable route, making it unusable for daily needs. Faster, more frequent bus services are accessible along the main valley road, but reaching the bus stop requires a 5 km walk with a 300 m elevation gain without a car.

In Palmira's case, another type of constrained immobility is observed, not due to her physical or cognitive abilities—she has significant self-management and mobility potential compared to the elderly interviewed—but due to the lack of certain requisites, such as owenership and ability to drive a private vehicle. This would be crucial for accessing off-site activities Palmira wishes to engage in more frequently, like personal care or visiting her sons living far from Cerignale. Palmira's immobility is thus partly experienced as a constraint, but this constraint is somewhat counterbalanced by her choice to live in a desired place (Palmira returned to Cerignale after living in other larger cities, aware of the difficulties she would have to face) and the support systems available locally (store, social taxi, community solidarity) that enable her to find a balance between unmet needs and those locally available, essential for her daily life.

5.3.2 Enablers of Immobility

The ability to live daily life while limiting personal mobility to the proximity of one's home, particularly in the Apennine high valleys, depends on either the physical proximity to necessary activities or alternative access methods that do not require physical movement, such as virtual means or support from others. This latter scenario involves more mobile individuals enabling the immobility of others through their care, assistance, and presence.

During fieldwork, various individuals exemplifying this role were identified. For instance, the case of Pamela, a resident of Cerignale who rarely leaves her apartment, illustrates how her immobility is supported by the mobilization of others, allowing her to access essential services related to health care, sustenance, and social interaction. The action of these immobility enablers can be linked to institutionalized activities under regional and local policies. The home-based socio-health care system for the elderly or those with reduced mobility, developed in recent years under the model of territorial and preventive medicine, is a prime example. In the Emilia Romagna region, territorial medicine is structured according to a logic of decentralization implemented by creating intermediate structures between the general practitioner and the hospital, the so-called *Case della salute* (health homes). Casa della salute is conceived as an outpatient center that provides ordinary care and screening and organizes and coordinates the activities of doctors and nurses who visit patients in small village clinics and patients' homes for home care.

Luciana, a nurse at the Casa della salute in Bettola, exemplifies this role. She travels daily from village to village, assisting many, especially the elderly, who require frequent care. Luciana not only provides healthcare but also acts as a support figure in fragmented communities where many older people live alone, the last inhabitants of small hamlets that are now abandoned or inhabited only for a few months of the year. The trust established with patients often extends to informal support, such as delivering groceries, which helps rebuild community ties and mitigates the exclusionary effects of constrained immobility.

Beyond welfare-based healthcare services, family members and privately paid caregivers also play crucial roles. In the cases of Pamela, Giovanni and Tiziana, caregivers represent an essential resource to lighten the burden of the relatives' care for their parents, especially during the night and in cases where their children live far from the villages. For example, Pamela's caregiver, Paula, from Latin America, lives with Pamela during the week, assisting with shopping and daily activities. Other people in the village share the same profession, creating a parallel immobile population not recorded at the registry office of the municipality. Paula lives in Cerignale during the week, and her sphere of activity is only slightly more extended than that of the person she takes care of. Only during weekends, on her free day, she is accompanied to the bus stop in the valley from which she reaches Piacenza to visit her cousin who lives there. Thus, unlike nurses, caregivers often immobilize themselves, constrained by their responsibilities and work choices rather than by the inability or lack of interest in moving, and adapt their routines to those they care for, regaining mobility only on their days off.

Immobility enablers also facilitate access to other types of goods and services in the territory. Barbara has been working since 2017 in the small grocery store in Cerignale that acts as a place for aggregation and sociality (Barbara has created a small bar next to the store where she serves homemade products) and that is perceived as a fundamental resource for the daily subsistence of the inhabitants. The store opened in 2017, in contrast to what happened in many other villages of the Piacenza Apennines where demographic decline and more attractive activities on the valley floor forced the closures of these kinds of shops. For the atypical nature of this opening, strongly desired and supported by the municipal administration of Cerignale, the store is seen as a valuable resource to be proud of and to preserve. Yet, although Barbara is considered as a member of the local community, she does not live in Cerignale but near Piacenza. After dropping her children off at school, she drives to the village and opens her store every morning. In the summer vacation period, when a larger number of customers is present in the village, Barbara extends her opening hours and takes her children to work, leaving them free to move about and interact with other young children of tourists or vacationers who are temporarily relocating. Barbara's mobility practices are therefore carried out daily over a very extended geographical area, requiring time and commitment in reconciling family and work activities, and represent a primary factor in ensuring that the inhabitants of Cerignale can satisfy some of their daily subsistence needs without having to move away from their sphere of proximity. The opportunity offered by Barbara was particularly significant during the period of the total lockdown in Italy linked to the Covid-19 outbreak. In that circumstance, Barbara continued to offer her support by creating a network of home deliveries for local inhabitants who would have had difficulty in accessing such goods given their constrained condition of immobility that, in that period, generally limited travel also for usually mobile people.

Similar resources are no longer available, at least not permanently in most of the other hamlets. However, interviews revealed the existence of several informal forms of accessibility provision. For example, valley floor pharmacists home-deliver medicines to patients in remote hamlets or entrust them to the nurses providing home healthcare. The same happens with other types of goods, with some itinerant traders who move from settlement to settlement with vans selling basic food and clothing goods. These services can be requested by telephone, but they are mostly based on custom: the traders pass through on defined days known to the inhabitants. These activities are difficult to identify and map as they result from a few spontaneous individuals' initiatives, mobilizing to reach clients and compensating for service deficiencies and limited participation among the most immobile inhabitants.

The role of immobility enablers is thus essential in the Apennines of Piacenza to ensure that constrained immobility, caused by individuals' inability and poor access to goods and services, can be partly reversed, at least for some essential needs. Additionally, the presence of these enablers can help foster a stronger sense of community among inhabitants who rely on local activities and build relationships that go beyond mere commercial transactions or healthcare services. In this context, the ability to physically access a nearby service is just one factor in reversing constrained immobility, as it is intertwined with a complex array of relational and emotional factors related to belonging to a socio-spatial context, which can modify one's perception of their condition, desires, and limits.

5.3.3   Immobiles by Choice

Being immobile in one's daily life and rarely moving outside one's ideal area of activity can sometimes be a deliberate choice. Claudio and Francesco, two brothers in their 40s, decided to settle permanently with their families on a restored old farm in Crocenito, along the mountain road between Val Nure and Val Trebbia, within Bettola's municipality. Reaching the nearest services, located in Bettola or in Perino, a hamlet in the municipality of Coli, requires a 20 min drive and overcoming a 350 m height difference, indicating their farm's isolation from accessible services as those considered in the accessibility calculation proposed in Chap. 3. The brothers chose this area for its isolation and agricultural potential, allowing them to live self-sufficiently by consuming what they produce and tending to their livestock and land. They also sell their products to visitors, especially on weekends.

The two brothers dedicate their daily lives to running the farm and rarely travel further afield. Their movements are mainly limited to Bettola, which they reach to carry out errands and bureaucratic activities related to the management of the farm or to buy subsistence goods that they do not produce directly. This relative immobility is a conscious life choice, aligning them with a growing group of amenity migrants (Moss 2006) or mountaineers by choice (Dematteis 2011). In fact, for Claudio and Francesco, self-sufficiency represents a reference value based on the awareness that many of the resources they need for their well-being is available in their close proximity. Their occasional trips to Bettola demonstrate that their immobility is a choice, not a constraint. Indeed, the two would have the capacity to access and participate in activities not found locally through more frequent and complex mobility. However, the ability is converted into functioning only if deemed necessary, and in any case infrequently, because many of the material and immaterial goods they need and desire are located within walking distance of their home. For this reason, even if the low physical accessibility to services and opportunities that characterize their living place would suggest considering them as constrained immobiles, the fact that they voluntarily choose this specific condition for the intangible benefits it provides suggests that it would be more appropriate to indicate this as a condition of 'chosen' reversible immobility. However, it is worth noting that forms of reversible immobility are subject to changes and reorientations linked to transformations and evolutions in the individual's life. At the time of the interview, Francesco was beginning to wonder how to organize his daily commute to and from Bettola to take his older son to kindergarten. In the absence of a transport system that reaches the hamlet, it becomes necessary to make frequent movements towards the main center. While remaining within the municipality of reference, these require at least two round trips, for about 80 km per day, redirecting the practices of immobility experienced by these people towards different forms induced by the dependence of other relatives towards the potentially more mobile members.

The farmhouse is also a meeting place for the nearby people who perceive it as a place for social relations. During a visit, Federico and his girlfriend were in the farm. They moved from Piacenza and Genoa to a nearby rural house for similar reasons as

Claudio and Francesco. Federico, a professional educator, plans to start a local home-schooling group to involve children of the inhabitants of the dispersed hamlets of the municipality of Bettola. This experience shows how some new inhabitants of the area have chosen to repopulate these places in search of particular conditions related to isolation and perceived distance from specific values and rhythms associated with the urban dimension (Dematteis 2018). Therefore, factors such as accessibility to services and opportunities offered by urban centers seem to have less weight on the choices and behaviors of the people interviewed. For them, propensity for self-sufficiency prevails, and reversible immobility, expressed as rootedness in a territory following criteria of slowness and proximity to the local community, becomes an identity value that these people might not satisfy by living in more dynamic and accessible contexts.

Cesare's story is different for some extent. Together with his family, he manages a hotel in the hamlet of Capannette di Pey, in Zerba. Capannette is located at 1400 m in altitude, on the border that links Emila Romagna, Piedmont and Lombardy. The locality is mainly inhabited in summer when it becomes a busy center for sports activities and hiking in the mountain environment. Even if it is part of the municipality of Zerba, Capannette is about 600 m in altitude and can be reached by driving for about 30 min along the provincial road that connects it to the main village. Cesare manages the refuge owned by his family where he was born and raised, and during the summer season he permanently moves to Capannette. He works primarily around the hotel in the summertime, receiving vendors, tending the surrounding grounds and managing hospitality with his family. When he rarely needs access to services that he does not find in Capannette and the restaurant's suppliers cannot provide (e.g., access to the pharmacy or the doctor), Cesare uses his car to drive to Varzi, 50 min further north in Lombardy. Cesare and his family prefer to go to Varzi because they find the offer of services more complete than those available in Ottone, the first settlement of certain importance in Val Trebbia, located at the same distance from Capannette. Moreover, unlike Francesco and Claudio, who permanently reside on their farm, Cesare does not spend the whole year in Capannette: at the end of the summer season, around September or October, Cesare moves to Novi Ligure, an hour and 15 min by car in the flat areas of southern Piedmont, where he lives during the winter months to allow his children to attend school and other social activities. He moves back to Capannette in spring. The relative immobility of proximity that Cesare experiences in the summer period is thus interrupted by infrequent mobilities that extend over a significant area, and which are surprisingly unrelated to the administrative dimension: Cesare lives in Zerba in Emilia-Romagna for a few months, during which he sporadically uses the services of Varzi, in Lombardy. However, he lives in Novi Ligure, in Piedmont, for the rest of the year. Cesare's multiresidentiality describes an intermittence between moments of relative immobility interspersed with wider movements that follow a kind of seasonality and respond to Cesare's need to reconcile his working life, his roots in the place where he was born, and the needs of his family.

The rooting and return to these territories seem to be the goal of the actions promoted by the municipality of Cerignale, taking advantage of the Covid-19 pandemic effects. During the interview, the village mayor talked about a marked phenomenon of return to the village following the total lockdown (March–May 2020) by occasional tourists and vacationers who usually attend Cerignale only for a few weeks in the summer period. The contingency determined by the limited opening of schools and the transfer of many work and study activities in remote forms, even after the first phase of reopening, has led the village to come alive well in advance of previous years and with a more significant presence of people. To make up for the lack of performing digital connections and the absence of wired spaces in which to work, the municipality has created coworking facility in the town hall and a multi-functional room in the small recently opened library. Thanks to the presence of a fixed public internet connection, the intervention aims to meet the needs expressed by a growing group of people interested in moving for longer or shorter periods in the small village, providing them with some proximity services that can support their daily activities while remaining in the village.

In perspective, these actions aim to limit the inconvenience and costs of the lack of physical accessibility to opportunities, focusing on the reorganization of work and the availability of technologies that enable possible forms of reversible immo-bility. Moreover, they aim to diversify the possible newcomers, attracting not only people interested in opportunities in the field of agriculture and land management, but also different professionals who can perform their activities in total or partial form remotely. These actions are implemented to reconstitute a local social and community tissue that would otherwise risk being lost.

5.3.4 Mobile People, Between Constraint and Choice

The interviews conducted in the area provided insights into various forms, methods, and consequences of social participation related to relative immobility in daily prac-tices. As discussed, some forms of immobility can be traced back to a choice if read concerning the effects they can generate on the possibilities of physical access to goods and services and the individual's perception of inclusion and wellbeing. In other cases, however, they may result from circumstances that prevent one from moving as much as one would like. However, it is interesting to note that the bound-aries between choice and constraint are never entirely clear: the same individual may perceive greater degrees of dissatisfaction with the impossibility of achieving some opportunities and less for others. The boundaries between choice and constraint are just as blurred when collecting the stories of people who, for various reasons, are highly mobile in their daily practices.

Sara is from Zerba, a municipality with limited availability and access to essential services and absence of public transportation. Sara and her husband have lived in the village for many years and have both been able to find a local employment opportunity. Their daily lives necessitate traveling by car to reach the nearest services in Ottone, a village 15 min away by car, with a 330 m elevation difference and a two-hour walking distance. During winter, snow-covered streets can make even car travel difficult, extending driving times.

During the Covid-19 lockdown, younger adults in the village took turns shopping for the entire community, delivering food to the elderly, fostering mutual support based on relational proximity and enabling immobility. For Zerba's residents, reaching Ottone also means accessing the closest primary and middle schools, pharmacy, and public transport, all of which are absent in Zerba. At the moment of the survey, Sara's son had just started high school in Bobbio, which is the closest to Zerba but still about 40 min by car or 80 min by bus. Sara drives her son to Ottone every morning before returning to work in Zerba. Her son will then take the bus on the valley floor to Bobbio. Sara noted that living in a remote village often prompts residential migration for high school education, especially for families without strong local ties. Her son chose a specialized institute in Bobbio, but if he had opted for a school in Piacenza, the family would have faced different strategies to manage commuting or considered options like student residences. Families with more than one child attending schools in different villages encounter additional challenges in reconciling travel and family schedules. Sara also highlighted the difficulties in managing her son's social relationships, as he must adapt to public transport schedules and his parents' routines due to his lack of autonomy. However, the trips to accompany their son to the bus stop are used by Sara and her husband to run errands and shop in Ottone, which are necessary due to the absence of stores and proximity services in Zerba.

Sara's experience underscores that despite working close to home, complex and extensive mobility is required to ensure that non-self-sufficient family and community members can access goods and services. Furthermore, it also highlights how younger residents, who are almost absent in statistical data, emerge as category with lower accessibility to their basic needs, as the school, sport, social and leisure activities are located in the major villages of the valley floor, thus inaccessible without a car. The lack of opportunities for youngsters is a relevant issue that risks being underestimated by only considering statistical data and the resulting in an underrepresentation of these social groups.

Antonio, an older man living alone not far from Claudio and Francesco's farmhouse, provides another perspective on mobility and immobility. He lives in an old building where he has always lived and cultivates the land. After an accident left him without a car for an extended period, he used his tractor for mobility to avoid staying home. Compared to other interviewees of his age, Antonio remains very mobile and independent. He frequently travels to Bettola, about 20 min away by car, for daily errands and social interaction. He also frequently visits the nearby farmhouse of Claudio and Francesco, especially during the lockdown seeking company and support during the months of isolation. Although eligible, Antonio requested home care only for a limited period following his accident and refused assistance after his recovery. He expresses a strong determination to remain mobile, even stating he would walk to Bettola if his driving license were not renewed. Antonio is deeply attached to his home despite recognizing that life is becoming increasingly complex: he needs to move but finds it increasingly difficult to do so as much as he would like. Moreover, he has no children and cannot count on daily assistance, which is partially guaranteed by the proximity of the farmhouse. For these reasons, he has thought of renting an apartment in the center of Bettola to be closer to stores, services and people he knows, at least during winter months when weather conditions further complicate mobility.

Loredana and her husband live in the small village of Costa Rodi, a hamlet in the municipal territory of Bettola, although located 40 min driving far from the center. Loredana grew up in Costa Rodi but moved to Piacenza to attend secondary school and work, following a typical migration path due to educational reasons from the mountain to the valley. In the last few years, Loredana and her husband have often used the house in Costa Rodi for a few weeks during the summer period. Now, with her husband's retirement, they split their time between Piacenza and Costa Rodi's family house, spending six months in each location. Since she is still working in Piacenza, when in Costa Rodi she commutes 45 min by car daily to reach her job place and return. The fact that she travels daily to Piacenza also allows her to do activities that she finds along the road or near her workplace, which are absent near Costa Rodi. Hence, car ownership becomes crucial for residents to access daily activities and essential services, as confirmed by accessibility analysis results. Indeed, despite the availability of a demand-responsive transport (DRT) service connecting Costa Rodi to the centre of the municipality in Bettola, this is a mobility option that is considered impractical due to distance to the transit stops, travel time and low frequency. Costa Rodi's residents prefer traveling to Coli for essential services, utilizing a shuttle bus managed by the municipality that brings people to the local market on Mondays, and which is perceived as more efficient and able to respond to specific accessibility needs as reaching the weekly market. Other opportunities are also offered by the periodic passage of some salespersons who sell frozen food at home and other necessities to the few who still live permanently in the hamlet.

In contrast to Loredana's partial return to the Apennine areas, Raffaele chose the opposite path. After reaching retirement age, he dedicated himself to beekeeping, turning his passion into a profession. Several times a week, he travels from Piacenza, where he lives with his wife, to Groppavisdomo, a hamlet in the municipality of Gropparello, using an old family house for his beekeeping activities. On weekends, he extends his mobility by traveling throughout the province to sell his products at farmers' markets. Raffaele and his wife decided to remain in Piacenza due to proximity to services and social connections. However, their son, also a beekeeping enthusiast, recently moved from the city with his family to a farmhouse in the mountains to start a honey production company. Although elderly, Raffaele remains more mobile daily than his son, commuting almost daily between Piacenza and Groppavisdomo, while his son is now rooted in the Apennine area.

The mobile people interviewed express these practices for different reasons and intensities and could shift towards reversible immobility if circumstances change, such as Sara's son gaining autonomy in his displacements or Antonio deciding to relocate to Bettola. Similarly, Loredana might choose to stay more permanently in Costa Rodi after retirement, or Raffaele might find beekeeping too burdensome to continue.

All the mobility practices analyzed are somewhat ambiguous if read in relation to how the individual may feel more or less compelled to carry them out given the necessity to move over long distances to satisfy needs that, in this territory, usually cannot find their fulfillment in proximity. In the case of Sara, the mobility practices she carries out daily are influenced, in their complexity and extension, by spatial conditions and distance from local and territorial services, requiring her to dedicate significant time and effort to traveling through the valleys and up to Zerba. Loredana's situation differs: despite the option to live closer to her job in Piacenza, she chooses to travel long distances daily for part of the year. Instead, Raffaele moves away from Piacenza to carry out an activity that cannot take place near where he lives in the city, moved mainly by passion. Finally, Antonio evaluates the hypothesis of finding a compromise between residential location and accessibility in the perspective of not being able to maintain for much longer his currently available resources and capabilities for mobility.

At least two conditions unite these people, especially those who reside more or less temporarily in the Apennine areas (Antonio, Loredana, Sara). First, they all face the absence of activities and opportunities to which they respond through their frequent and complex car-based mobilities, which often find little response in the opportunities offered by local public transport systems. Second, their mobility, and the trips they must take, are the expression of rootedness in the place where they live more or less temporarily: in their cases, and according to their needs and preferences, only being mobile guarantees the possibility to stay put in the places to which, for different reasons, they are linked.

5.4 Different Profiles: Systematizing Reversibility Factors

The interviews conducted in the Apennine area of the province of Piacenza have returned a complex, varied, and above all dynamic picture of the mobility and immobility practices of inhabitants that, rather than being static in time and space, end up being made up of relationships, interdependencies, and changing preferences as described for each profile (Table 5.1).

In an area with low accessibility to services and opportunities, where forms of mobility and immobility would be considered constrained according to the conceptual scheme proposed in Chap. 2, the stories of the inhabitants and users of the territory collected during the fieldwork have shown how complicated it can be to profile forms of prevalent mobility and immobility and to evaluate the implications in terms of social participation. Every single experience describes how individuals can be simultaneously enabled and constrained by the practices of daily immobility they perform: they can, for example, draw satisfaction from participating in the available material and immaterial opportunities they can access in proximity, or from those remote and challenging to access that they can reach through long trips at a greater or lesser frequency. In other cases, capabilities will drive them to both use some local opportunities and, if they can, move to reach others not available locally or deemed more appropriate to fulfill their needs. It is also common that a condition in which certain mobility or immobility-oriented behaviors prevail can reverse in time

Table 5.1 Synthesis of the emerging profiles

Constrained immobiles	Immobility enablers	Immobiles by choice	Mobiles
Can rely on **limited personal motility** (age, health, physical autonomy) carrying out their daily activities in a very restricted space, in line with their limited resources and capabilities For some, this downsizing may not be always constrained based on **adaptive preferences if other resources are in place allowing their immobility** For others, the constrained condition **may be due to the lack of requisites, abilities and resources or interdependence** In many cases, **other immaterial factors** (i.e. sense of community) can contribute reversing their apparently constrained condition	More or less mobile people who become **enablers of immobility for other inhabitants providing care, assistance, and presence** Examples include medical staff providing home care (institutional assistance), caregivers, shopowners, salespersons **They are essential to ensure that some forms of immobility constrained by people's poor accessibility is made reversible** Their activity generates a complex series of emotional factors linked to sense of community that can, in turn, modify the perception of one's condition, one's desires and one's limits	Their immobility can **be the result of a choice to which specific values and rationalities are attached** For some, a limited activity area is the result of the desire for rooting, self-sufficiency and isolation: **many of the material and immaterial goods they need and desire are located within walking distance of their home** However, it is worth noting that forms of reversible immobility are **subject to changes** and reorientations linked to seasonality, changes and evolutions in the individual's life New forms of local rooting are also promoted through policies	Highly mobile people practices are **ambiguous to interpret** if read in relation to how they feel forced or not to carry them out given the necessity to move over long distances to satisfy needs that usually cannot find their fulfilment in proximity **They all face the absence of activities and opportunities to which they respond through their frequent and complex car-based mobilities** On the other hand, their mobilities are **the expression of rootedness** in the place where they live more or less temporarily: only being mobile guarantees the possibility to stay put in the places to which, for different reasons, they are linked

based on changes in one's preferences, needs, relations with others, and contextual conditions. For example, interviews have identified how increased direct and indirect access to opportunities in proximity, available mobility options and local relational resources, often induced or influenced by the effects of local policies (i.e., home care assistance, social taxi) act as a powerful reverser of otherwise constrained conditions.

In this perspective, and in line with the experiences gathered during the field work, one can identify the existence of *reversibility factors* that act on the actual, potential, and perceived personal possibilities of access and participation, making conditions of apparent or quantitatively measured constrained immobility perceived or experienced by the individual as reversible. Reversibility factors may largely vary in nature and effect and could be both positive (when fostering the perception of potential participation) and negative. Figure 5.2 shows some of these factors experienced directly by the inhabitants of the Piacenza Apennine context emerged from their microstories.

The previous scheme can, in turn, be generalized, as proposed in Fig. 5.3, extracting the factors that can play a positive role in the reversion from constrained to reversible forms of mobility and immobility.

The qualitative approach thus makes it possible to reconsider the conceptual framework proposed in Chap. 2, introducing a multiplicity of elements that make it more suitable for describing the dynamism and the relevant role of reversibility factors in the often-ambiguous relationships between forms of immobility at the individual personal level. Furthermore, the updated scheme orients general integrated transport and land use planning policies and actions that may induce possible reversion of constrained mobilities and immobilities related to limited accessibility to opportunities, as deepened in the next chapter.

Looking in-depth at the reversibility factors that emerged, Figs. 5.2 and 5.3 show that some individuals who may be relatively more immobile than others may be so due to their inability to move and access activities and opportunities outside their physical proximity in a direct way. However, they can also adapt to this constrained condition by taking advantage of alternative opportunities for indirect access (e.g., home care, delivery, digital resources), reducing their preferences (adaptive preference), or drawing satisfaction and well-being from resources, including immaterial ones, that they can access in proximity. From a policy-oriented perspective in the field of transport and land use planning, these forms of constrained immobility can be reversed by *immobility enablement* through the promotion of accessibility by proximity, both directly, by investing in the improvement and diffusion of local services and opportunities, and indirectly, through a combination of activities provided at home or in digital format.

Forms of constrained immobility can also be reversed by *mobility enablement*, namely creating the conditions for an immobile person to facilitate their physical displacements, ameliorating the connections with places where they can find the resources they want or need to access through mobility and travel since not available and cannot be provided locally. At the same time, those who are immobile by choice while having the ability to move and access goods and services, even if not spatially close, may reverse their daily relative immobility into mobility to respond to changes in their needs and desires (accessing activities and opportunities that are more remote or alternative to those available) or to allow others who are not autonomous in their movements to access resources they want or need, creating interdependencies. From a policy perspective, improving available mobility services could hypothetically facilitate the transition to—or alternation between—forms of reversible immobility and mobility. Conversely, investing in enhancing available opportunities by fostering direct and indirect proximity access to activities and services could induce an adjustment of preferences limiting the need for travel and favoring reversible forms of immobility.

People living in conditions of constrained mobility who must frequently move over long distances in conditions of low accessibility and availability of proximate

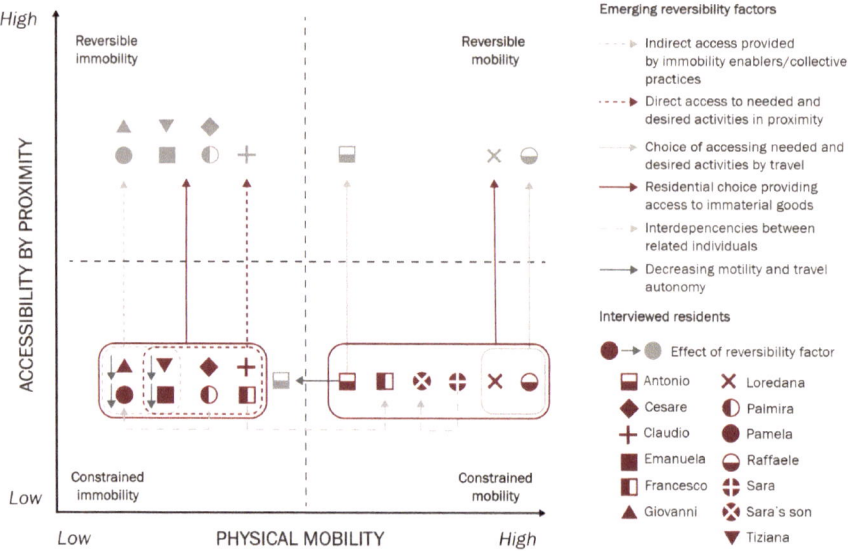

Fig. 5.2 Reversibility factors emerged from the interviews with local residents

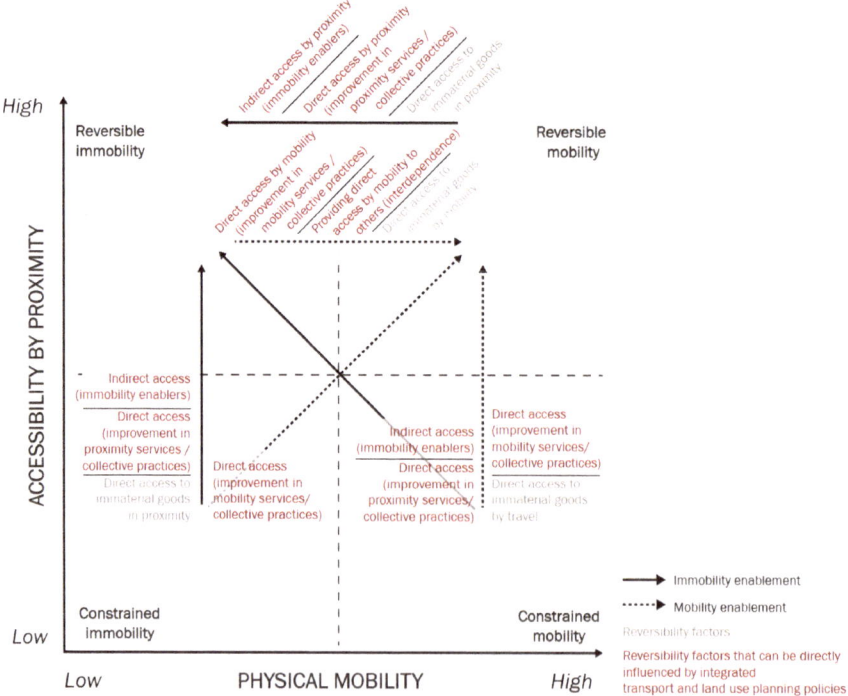

Fig. 5.3 Generalized account of the observed positive reversibility factors

activities and services may, in turn, perform these mobility practices for reasons more or less oriented by their own choices and preferences and constrained by the spatial conditions of the contexts in which they live and move. In some cases, traveling frequently can be considered a necessary burden that one would do without, while, for others, it may be a choice that one decides to make even in the presence of less travel-demanding alternatives because they are perceived as less preferable for one's personal satisfaction and well-being. Considering possible integrated transport and land use policies, improved mobility services can facilitate the necessary travel towards distant resources and activities, possibly redirecting them towards more sustainable forms. In parallel, investment in local development favoring direct and indirect accessibility by proximity and immobility enablement may help reverse the possible constraints linked to highly mobile behaviors.

Despite the variety and ambiguity of the different circumstances that emerged and the possible inter sectorial policies and measures they can inspire, accessibility by proximity appears as a key element in shaping the forms of immobility of the interviewees and qualifying them according to the level of participation in the activities that these forms guarantee. This participation, however, is not only given by the possibility of physically accessing the desired activities in proximity or at a distance with greater or lesser autonomy, but also by the result of specific actions based on interdependence among community members. For instance, while some institutional immobility enablers allow local people to indirectly access specific welfare services (e.g., home care), there is a proliferation of informal and spontaneous opportunities supporting access and participation in the territory, providing both mobility and immobility enablement. Indeed, the interdependence among community members leads to the emergence of recurring practices based on spontaneous cooperation, support, and self-organization and management. These collective practices challenge the definition of accessibility proposed in Chap. 2 as the ability of an individual to participate in needed or desired spatially distributed activities based on individual and contextual factors. Instead, in this perspective, accessibility can also be viewed as a collective good provided by actions performed by a group and not just resulting from individual efforts. Consequently, the capabilities an individual may lack, which could limit their mobility and accessibility, can be supplemented by other community members, creating opportunities for mutual support based on interdependent relationships. These collective practices both providing local access in proximity to community members or supporting the mobility needs of others were observed in the context of the Piacenza Apennines and were frequently cited by residents as significant reversibility factors. Examples include receiving a ride from a neighbour, spontaneous actions by pharmacists delivering medications, and the creation of local interaction spaces of proximity in small villages (e.g., the grocery store or the remote working room in Cerignale), all of which offer multiple opportunities for mobility and immobility enablement and reversibility from constrained conditions, as shown in Fig. 5.3. These micro-practices may be partially supported and recognized within institutional accessibility provisions from local authorities (e.g., transport providers) and may overlap with their scope, even generating possible competition. However, their role is essential in addressing specific deficiencies in institutionalized accessibility provision, such as the absence of an efficient and reliable public transport system and the scarcity of welfare services in proximity. Tracing these practices based on relational more than physical proximity to goods and services is certainly challenging for researchers and planners, but it provides relevant insights as community solutions often respond adequately and innovatively to local accessibility needs, potentially offering important elements for designing and adapting policy measures for mobility and immobility enablement based on local knowledge and needs.

Related to this point, a significant (though difficult to identify and measure) role is given by the satisfaction that people derive from their immobility practices and the awareness of having access to immaterial goods available in their living context. For instance, the conditions that make a person happy to reside in a place they feel connected to, or the possibility of sharing that place with a community where emotional bonds may be established, emerge as values able to counterbalance, at least partially, the limited possibilities of physical access for some respondents. Indeed, access to these immaterial goods represents a significant factor of reversibility between constrained and chosen immobilities, besides being a resource for inclusion and participation of people living in peripheral and marginal contexts. If, on the one hand, these intangible factors depend largely on the expectations and needs of the individual, as well as preferences that change and adapt in different circumstances of life—therefore difficult to be assumed from a policy perspective—it is possible to imagine that actions to improve accessibility and enablement of reversible forms of mobility and immobility can still have an impact on this sphere, even as a side effect of the same policies and measures.

5.5 Conclusions

Compared to an aggregate and static reading of the features inducing immobility described in the previous chapters of the book, the qualitative approach has allowed for more granular analysis resulting in an articulated and complex picture of the territory. The interviews revealed how people with similar backgrounds and social conditions live and experience immobility differently. Although it is possible to identify respondents based on the intensity of daily mobility (i.e., considering the spatial and temporal extent of their movements related to access to opportunities and services), the differentials that emerge between different individuals should not be interpreted as static over time. They change dynamically in response to various life circumstances and contextual variation in access needs, in fact contributing to reverse the conditions of the individual, as do possible policy measures and actions designed to achieve the same purpose. At the same time, it emerged how the relative immobilities of some are nothing more than micro-mobilities, expressed mainly in restricted spatial environments (either by choice or by constraint, or by a mix of the two conditions), underlining how an over-individual scale view can easily overlook the effect of these micro-practices that emerged as a critical qualifying element for the well-being of the interviewees.

Specific reversibility factors have emerged, based primarily on promoting direct or indirect proximity accessibility through the provision of local services and activities (immobility enablement) or facilitating direct access through mobility and travel to distant goods and services (mobility enablement). The role of communities has also emerged as a key factor since, through forms of interdependence and collaboration, it can create conditions for reversible mobility and immobility, conceiving accessibility as a collective rather than individual matter. Additionally, specific immaterial goods were highlighted as factors of reversibility, influencing the perception of living contexts based on the diverse opportunities they offer, which may be not only related to the physical presence and proximity to welfare services.

These insights contribute significantly to quantitative immobility and accessibility analysis tools and the design of more context-sensitive policy responses.

Firstly, the variety of experiences and needs revealed by the microstories confirm the discussion in Chap. 3, suggesting that identifying a univocal a priori set of needed activities and ways to reach them for aggregated populations to be included in accessibility assessment may be misleading and may drive ineffective policies. Instead, the list of opportunities that emerge from the interviews may indicate, for some places and some inhabitants, those activities to which it is a priority to ensure direct access by facilitating the mobility of people or improving the availability of the same activities in proximity. While it is impossible to design policies based on individual experience, a deeper knowledge of these dynamics may help better understand the reasons for marginality and design more inclusive territories, as Vecchio (2020) reported.

Furthermore, direct observation and interviews allow for overcoming traditional boundaries between disciplines. For example, welfare and health policies, even though not directly related to transport and mobility, may significantly affect accessibility levels. The same techniques, especially if framed in structured and institutionalised participatory processes, allow discovering bottom-up practices and self-serving networks, whose impacts are difficult to measure through more aggregated data and simulate in accessibility measurement. Nonetheless, these policies and practices shorten the distances between people and services and generate new forms of indirect accessibility contrasting marginality. By making these 'hidden' dynamics visible, this method enriches the representation of low dense, scattered, marginal territories provided by more traditional accessibility and mobility analysis.

Still, the use of qualitative data allows investigating the complexity of immobility practices in their spatial expression, both in larger areas and within the sub-municipal dimension. As explained in Chap. 3, mobility practices happening within a single municipal area are not captured through mobile phone data-based analysis but are still extremely common among inhabitants. Such local daily mobility practices are significant in terms of commitment and costs by people performing them and the access they provide to a range of services and activities concentrated only in the central cores of the municipalities. Given the extension of the municipal territories, movements within the same area can be particularly long and difficult. This condition leads one to question the idea that low variations in presence within a municipality can always be an expression of low mobility, as proposed in the methodology proposed in Chap. 3. Such a criterion can be used in dense urban areas, where information is collected following a subdivision of the territory into smaller sub-spaces but is potentially misleading if applied in dispersed and broad contexts with low settlement density.

Moreover, even if accessibility thresholds are generally established and measured based on municipal and provincial borders, the interviews reveal that those boundaries are not always observed when looking for essential services and performing daily mobility practices. These dynamics suggest the existence of an administrative mismatch in measuring accessibility.

If the qualitative approach tested in this work introduced new elements for improving accessibility assessment and measurement tools and for the progression of the debate regarding immobility and its social and spatial forms and implications, it should also be noticed that this approach has its limits.

First, it is time and money-consuming for the researcher. Then, as discussed, because of the sample's limited dimension, the information collected is not representative of the local population, and more structured engagement processes would be needed to gather policy-relevant insights. However, the low density of the area and the average population age also influenced the number of respondents, while the collected information may be considered a coherent representation of the needs and capacities of a highly represented part of the local population, namely the elderly, that, as it will be seen in the next chapter, are the primary recipient of most of the immobility enablement policies implemented in the territory. Thus, improvement may be made to the criteria of selection of the interviewees, which are based on personal connections established by interviewers, for example, including local young residents that, as they are underrepresented in demographic terms, could be considered a potentially 'forgotten' target for local public policies while suffering the effect of limited autonomy in accessing resources as experienced by other social groups.

References

Bissell D (2018) Transit life: how commuting is transforming our cities. In: Transit life: how commuting is transforming our cities. https://doi.org/10.1080/2325548x.2019.1579568

Cao M, Hickman R (2019) Understanding travel and differential capabilities and functionings in Beijing. Transp Policy 83:46–56. https://doi.org/10.1016/j.tranpol.2019.08.006

Dematteis G (2011) La metro-montagna: una città del futuro. In: P Bonora (ed) Visioni politiche del territorio. Per una nuova alleanza tra urbano e rurale, Archetipolibri, Bologna, pp 85–92

Dematteis G (2018) La metro-montagna di fronte alle sfide globali. Riflessioni a partire dal caso di Torino. Journal of Alpine Research|Revue de géographie alpine [En ligne], 106-2|2018, mis en ligne le 12 août 2018, consulté le 03 janvier 2022. http://journals.openedition.org/rga/4318; https://doi.org/10.4000/rga.4318

Goodson IF, Gill SR (2011) The narrative turn in social research. counterpoints, pp 386, 17–33. http://www.jstor.org/stable/42981362

Manzini E (2021) Abitare la prossimità. Idee per la città dei 15 minuti. Egea, Milano

Moss LAG (2006) The amenity migrants: seeking and sustaining mountains and their cultures. In: The amenity migrants: seeking and sustaining mountains and their cultures (Issue August 2006). https://doi.org/10.1659/mrd.mm008

Nussbaum M (2003) Capabilities as fundamental entitlements: Sen and social justice. Fem Econ 9(2–3):33–59. https://doi.org/10.1080/1354570022000077926

Ojermark A (2007) Presenting life histories: a literature review and annotated bibliography. Annica CPRC Working Paper 101

Polkinghorne DE (1995) Narrative configuration in qualitative analysis. Int J Qual Stud Educ 8(1):5–23. https://doi.org/10.1080/0951839950080103

Sen AK (2005) Human rights and capabilities. J Hum Dev 6(2):151–166. https://doi.org/10.1080/14649880500120491

Vecchio G (2020) Microstories of everyday mobilities and opportunities in Bogotá: a tool for bringing capabilities into urban mobility planning. J Transp Geogr 83(January):102652. https://doi.org/10.1016/j.jtrangeo.2020.102652

Vecchio G, Martens K (2021) Accessibility and the capabilities approach: a review of the literature and proposal for conceptual advancements. Transp Rev 1–22. https://doi.org/10.1080/01441647.2021.1931551

Vendemmia B, Lanza G (2022) Redefining marginality on Italian Apennines: an approach to reconsider the notion of basic needs in low density territories. REGION 9(2):131–148. https://doi.org/10.18335/region.v9i2.430

Verlinghieri E, Schwanen T (2020) Transport and mobility justice: evolving discussions. J Transp Geogr 87:102798. https://doi.org/10.1016/j.jtrangeo.2020.102798

Chapter 6
Public Policies for Immobility and Mobility Enablement

Abstract The concluding chapter of the book focuses on the promotion of chosen and reversible immobility and mobility through integrated planning and transport policies and is organized into two sections. The first addresses policies for immobility enablement, aimed at enhancing the direct or indirect provision of activities and services to reduce travel needs by fostering accessibility by proximity. The discussion draws on the recent debate surrounding the concept of the x-minute city and its practical applications in various global contexts, highlighting both potentials and limitations of these approaches. The second section reviews best practices for mobility enablement, exploring alternatives to traditional public transport supporting sustainable and shared mobility, facilitated by emerging technologies. Both sections reference policies applicable across diverse settlement contexts, extending models and visions traditionally associated with compact cities to encompass low-density peri-urban areas, ultimately providing a comprehensive overview of relevant planning policies for immobility and mobility enablement.

6.1 Introduction

The methodological approach developed in this book has elucidated the complexities inherent in analyzing, evaluating, and reconstructing differentials in mobility and relative immobility, particularly in their spatial and temporal dimensions. Moreover, it has highlighted the challenges, and, in some cases, the potential misconceptions involved in assuming that specific social, demographic or territorial conditions will produce clear mobility and immobility outcomes. Both direct and indirect access to services and opportunities—shaped not only by the physical and functional structure of the territory and the available modes of transportation, but also by individual preferences, needs, and perceptions—have emerged from the scientific literature and field interviews as factors influencing mobility practices and their implications for wellbeing and social inclusion.

These dynamics have been distinctly observed within the context of the Piacenza Apennines, which serves as the application field for the methodologies and analytical tools proposed in the preceding chapters of this book. This case study exemplifies the territorial situations typical of the so-called Global North, which can be associated with the phenomenon of peri-urbanism. Such areas should not merely be considered as transitional zones between town and country, but rather as diverse contexts where the multiplicities of peri-urban life unfold (Soja 2011 and Dodier 2013 in Pucci (2017)). These multiplicities reflect the relationship between individuals and communities and a regionally urbanized territory where activities and services are dispersed across space, population density is medium to low, and public transportation systems are relatively inefficient. The combination of these factors characterizes these contexts, notably in terms of the possibility that car dependency becomes an intrinsic and necessary aspect of daily mobility practices, thereby leading to exclusion from essential daily services for those without access to a car.

This scenario includes so-called rural areas, urban peripheries, urban outskirts, and suburbs—all of which are less connected and do not enjoy the same accessibility to activities and opportunities found in dense urban cores, thus qualifying them as peri-urban. In similar contexts of low accessibility by proximity, the immobility of people can, under certain conditions, represent a potential limit to social and activity participation. This condition happens, for example, in cases where accessibility by proximity to some relevant opportunities is not granted, and the individual does not have the capacity to reach those needed through their displacements or cannot rely on the assistance of others. Similarly, forms of disadvantage can also be mobility-related, when the needs to access distant resources and activities require costly and complex forms of high mobility, which may represent a burden to the individual. Such situations have been widely observed in the peri-urban context of the Piacenza Apennines. However, it should not be assumed that these issues are confined to peri-urban areas; consider, for example, the daily experience of individuals with limited mobility due to disabilities, where the impacts are felt regardless of settlement conditions. While the negative implications of immobility and low accessibility by proximity are exacerbated in peri-urban contexts, similarly challenging situations can thus arise even in urban environments, despite the usual presence of more favorable conditions due to higher density of activities and the availability of more reliable public or shared transportation options, potentially supporting active mobility options.

Considering accessibility as an essential good, as discussed in Chap. 2, prompts a reflection on how the disadvantages stemming from constrained mobility and immobility in contexts with limited availability of accessible activities and mobility options might be mitigated through targeted planning policies. However, it is crucial to emphasize, as proposed in Chap. 5, that conditions of mobility and immobility perceived as constrained are not solely linked to the dimension of individual physical accessibility (whether direct or indirect) to goods, services, and resources. On the contrary, a significant role is played by two other elements: first, the satisfaction that individuals derive from accessing certain intangible goods that are difficult to map or quantify analytically but can explain why a person may cope with a condition otherwise deemed disadvantaged if evaluated only through spatial accessibility assessments to services and activities. Second, the specific forms of interdependence that arise from the establishment of relational ties within a community, based on mutual interest, trust, and identity, which may materialize in informal access to goods, services, and resources. In such circumstances, access to intangible and relational goods—typically ensured by proximity to others within the same community—becomes the primary element enabling diverse and non-canonical forms of spatial access, which are not solely guaranteed by state or market-provided accessibility and mobility.

The main consequence of these considerations is that focusing on both the functional nature of accessibility, involving spaces, activities, and transport networks, and the socio-relational and collective dimension of accessibility is essential for promoting an approach to reversible immobility and mobility through planning policies. This approach leverages the activation of a plurality of reversibility factors, including those identified in the previous chapter. On one hand, this approach can be advanced through immobility enablement, which involves promoting accessibility by proximity through targeted, integrated land use policies aimed at enhancing the availability and reachability of essential activities via active modes. Ideally, these interventions could encourage the activity participation of those with low motility levels. Still, they can also nudge a possible change of behavior by those who, usually mobile in their daily life, would find it advantageous to redefine their spatial sphere of activity by limiting its extension thanks to the new opportunities made available in proximity, thus enabling forms of reversible immobility. Additionally, a potential outcome of these policies could be the establishment of shared proximity spaces, where forms of collaboration within the community could take shape or be strengthened, fostering a relational perspective.

On the other hand, policies aimed at reinforcing accessibility through sustainable mobility to reach unavailable local opportunities, defined as mobility enablement, could be relevant in all situations, particularly in many low or very low-density peri-urban contexts where it may not be feasible to ensure generalized access in proximity to such resources. These policies would support necessary displacements to meet more diverse accessibility needs.

This chapter aims to provide an overview of immobility (Sect. 6.2) and mobility enablement (Sect. 6.3) policies that integrally concern urban and transport planning, urban design, and the technological and process innovations applicable to these spheres. As mentioned earlier, this review proposes references and best practices that can be applied to various territorial contexts, albeit with particular effectiveness and urgency in peri-urban areas. The objective is to outline the opportunities and limitations of these policies in achieving environmentally, socially, and economically sustainable development goals that prioritize a comprehensive reconsideration of the role of immobility in contemporary societies. Concluding remarks close the chapter.

6.2 Policies for Immobility Enablement

The primary reference to frame a discussion on immobility enablement policies revolves today around the concept of the so-called "x-minute city." This concept advocates for rethinking the functional organization of urban settlements through policies related to transport, urban, and welfare planning, aiming to promote accessibility by proximity via a fair spatial (re)distribution of activities essential for daily life. Physical proximity to these opportunities, defined within the model according to varying time thresholds, facilitates their reachability through active modes—or collective/shared transport for longer distances—and lays the groundwork for the spontaneous development of forms of low mobility, or reversible relative immobility. This model, central to the current theoretical and political planning debate, was recently advanced by Carlos Moreno for the city of Paris and, as described in Chap. 2, has assumed renewed significance in post-pandemic global cities. The principles underlying it are regarded as promising frameworks for the development of more sustainable, inclusive, and resilient cities (Lanza and Pucci 2022). Despite its current prominence, this is not a "new" concept; as illustrated by Khavarian-Garmsir et al. (2023), the x-minute city and its various interpretations (from the prevalent 15 min city of Paris to the 20 min neighborhoods of Melbourne, including variations such as the 10 min city, the 10 min neighborhood, and 30 min territories) represent the latest evolution in a long-standing approach that considers the city from the perspective of the neighborhood and the local community. This approach can be traced in its historical development through several exemplary cases, including Ebenezer Howard's plan for the Garden City (1898), the Regional Plan for New York (1929) by Clarence Perry based on the concept of the "Neighborhood Unit," the modern movement's approach to the functional city, the "Environmental Area" proposed by Colin Buchanan in the 1960s, the new urbanism vision for organic city development (Alexander 1987), and Transit Oriented Development in the 1990s (Banister 1995). It is noteworthy that each of these visions aimed to propose a rationalization of urban development and, especially, mobility flows. Underlying these models is the idea of counteracting the expansive and fragmenting effects of urban sprawl to enhance accessibility and reduce the need for mobility, thereby diminishing the social, economic, and—according to more recent sensibilities—environmental costs associated with these practices. These models can thus be seen as actively promoting relative, reversible immobility as a positive benefit for both individuals and society as a whole.

Although influenced by a long tradition, the x-minute city model has recently garnered significant attention, inspiring numerous urban strategies and visions that are now being effectively implemented in several cities (Büttner et al. 2024), adapted to specific local needs, thereby providing increasing insights into the opportunities and limitations associated with their implementation. It is also important to note that many of these strategies pertain primarily to compact and dense urban environments worldwide. In discussing the nature of these immobility enablement policies, possible adaptations will be discussed to allow their transferability to peri-urban areas, where these policies must necessarily be reconsidered based on to the specific conditions of lower density and accessibility by proximity.

The various experiences mapped by Büttner et al. (2024) allow for the identification of four main types of policies attributable to immobility enablement. The first set of policies focuses on functional mixité, densification, and the decentralization of services. This can be achieved indirectly by defining specific urban planning norms in local plans to promote functional variety in neighborhoods, for example by limiting rigid zoning according to a principle of functional indifference, or by incentivizing urban development to areas that are already dense and highly accessible. Such principles, for example, underpin the recently approved urban plan of Milan (PGT 2030, Comune di Milano 2019). Direct methods, on the other hand, may involve the redistribution of welfare services across the territory and the diversified use of the same structures for different purposes, avoiding monofunctionality. Examples include the "Les Cours Oasis" project in Paris, which involves the use of school courtyards outside of school hours as accessible gardens for all (Ville de Paris 2023). A similar approach can also be extended to indoor spaces (e.g., school buildings) to provide a suitable location for different activities outside of school hours involving different members of the local community. Centralizing services can also play an important role in dense urban contexts to promote a more diversified and efficient use of public buildings. Centralization efforts based on the location of multiple public services in the same building, such as those promoted by the Stadsgebouw 2.0 City of Ghent project (Stadsbouwmeester Gent 2020), facilitate access to these facilities, since these are strategically located in central areas of neighborhoods to enhance accessibility through active mobility for the local community. At the same time, the availability of shared spaces among different users can help strengthen social ties, fostering participation and exchange. Similar policies can also be important for lower-density contexts, where activities and opportunities in proximity may be absent or limited. In this perspective, promoting immobility enablement by improving the availability and direct/indirect accessibility to some essential services, also achieving these objectives by leveraging initiatives already spontaneously put in place by local communities, can play a key role in ensuring inclusion, participation, and livability in these areas. These policies may involve different lines of action, starting with healthcare, which access, in peri-urban areas, may be compromised by the dispersion and distance of settlements from healthcare facilities. In this perspective, the home-care system represents a relevant resource for all territorial contexts precisely because it meets the needs of patients with reduced mobility options and abilities regardless of the characteristics of their living settlements. However, home-care assistance, a fundamental opportunity for immobility enablement, could be the starting point for an extension of the principle of indirect access to healthcare that could be potentially expanded to other age groups living in remote and low accessibility territories. Also, the opportunity offered through the provision of ICT-based healthcare services in situations where the health professional and the patient (or two professionals) are not in the same location (telemedicine) may represent a substantial asset for low accessible territories, as already widely experimented with within many European countries (Masella and Sgarbossa 2021). Thanks to its characteristics, and provided that the infrastructure of local digital connection allows it, telemedicine can become a key element in facilitating access to care and services even remotely. Also, education represents another important issue since, in low-density area, young people living outside the main centers may easily find themselves having to travel long distances daily to access school, and they also must adapt their rhythms and social interactions to the limited possibilities provided by the local public transport service and to the needs of the parents on whom they depend. These complex interdependencies represent a considerable cost for both children and parents, determined by the limited accessibility to schools and other local activities. Also, in more rural contexts, schools may be still physically present and fairly diffused to respond to the needs of a larger number of inhabitants from the past but increasingly threatened by the progressive depopulation and the lack of young pupils.

From a strategic perspective, these conditions of—at least partial- constrained mobility could be reversed by integrating classical school activities with new digital technologies that can support forms of remote learning, thus avoiding complex and expensive travel for students, as discussed in Chap. 2. Remote learning could also be coupled with forms of virtual collaboration between local schools and the promotion of itinerant learning infrastructures. This could be a suitable option both to keep school activities functioning in case of unexpected events (as happened during the Covid-19 Pandemic) and to meet the challenges faced daily by pupils and families with difficulties in physically accessing schools, as shown with success in some experimentations such as the Scuola@Bardi project in the village of Bardi, Italian Apennines, started about twenty years ago that allows young people of the municipality of Bardi (Parma) to follow a partially digital learning path in the local school avoiding long trips to other centers. Also, investing in functional mixite could be foreseen, for example by developing baby and after-school care in the already existing schools or other public facilities to ease the reconciliation of work and family care and reconfigure forms of interdependency between children and their parents and guardians. Also, the establishment of new opportunities for professional and entrepreneurial training for young people and adults could be foreseen, for instance, by envisaging the creation of local schools for the advanced teaching and transmission of the know-how of territory-related professions and trades, especially in rural and mountain areas: creating new services for good quality learning makes it possible for young people to root in the territory at the end of their studies and limit outward mobility towards more dynamic areas of the country.

To make the proposed interventions concretely possible and more effective, a strategic action could concern the realization of 'community hubs' to be located in places currently lacking in proximity services but easily accessible by foot and both public and private motorized means from the most peripheral areas of the territory where the establishment of these services cannot be guaranteed. This proposal could take concrete form in the identification of possibly disused or underused spaces in which to design the hubs, connecting them to the broadband network—and where to concentrate the offer of services and spaces for the community, such as territorial medicine activities, coworking, locations to access and perform distance learning, baby care and after-school, with a view to efficient and fair territorial distribution, as proposed in different experience such as the Dorpspunt (Village points) in rural areas in Belgium (Dorpspunt project in Beveren 2024).

A second set of policies concerns the redistribution of public spaces to facilitate local mobility. As discussed above, immobility is a relative concept that is based on the evaluation of the extension in time and space of one's mobility patterns. Reversible immobility is relative, since it contemplates a reduction in time and space—not the full absence—of travel to reach services and activities, which can be supported by a whole series of sustainable micromobilities of proximity. These flows, essential for ensuring access to neighborhood services, are supported in the x-minute city model, by interventions aimed at improving pedestrian and cycling connections. These interventions involve altering the form and use of streets to make them safer, more welcoming, and conducive to social interaction. This approach moves beyond viewing streets merely as channels for mobility or places where to park vehicles. Instead, streets are reimagined as the quintessential sites of human interaction (Gehl 2011). The interventions promoted by cities, which can also be applied to peri-urban contexts, are thus oriented towards reclaiming public spaces that have been predominantly used for non-social purposes. This is achieved through pedestrianization efforts, traffic calming measures, enhancements to urban furniture to create convivial and multifunctional spaces for diverse populations, and the securing of pathways. The goals are twofold: in the short term, to improve the overall quality of spaces and routes, thereby positively impacting local accessibility and neighborhood livability; in the medium to long term, to establish a behavioral change focused on proximity and active mobility, promoting reversible forms of immobility. In this context, redesign interventions that make streets "for people and not for traffic," even if conducted experimentally, can consolidate a broader transition towards different urban mobility practices (Bertolini 2020), even in peri-urban settings. The actions undertaken by various cities align with these principles, starting with the reference model of Barcelona's superillas (Ajuntament de Barcelona 2014), where traffic rationalization and the reorganization of spaces towards a pedestrian-oriented approach have laid the groundwork for a comprehensive improvement in the quality of public spaces, simultaneously enhancing access to neighborhood services and economic activities.

The impact of street transformations on the economy directly connects to the third category of policies aimed at promoting local economic activities. In dense urban areas, interventions that redesign and repurpose neighborhood streets can be accompanied by the promotion of production and consumption practices aligned with the principle of localism, as exemplified by Paris's "Fabriqué à Paris," an initiative providing economic support to local production activities (Ville de Paris 2022), or projects promoting the local consumption of short supply chain agricultural products. These initiatives aim to revitalize the neighborhood economy and the social fabric, generating local connections that could lead to reduced mobility needs for residents, both as users and employees of these locally significant services. This type of initiative can be particularly relevant in economically less dynamic areas, such as various peri-urban regions. For example, the proposal for community hub could be integrated with other forms of local development, for instance related to the investment on commercial activities that, as shown in some villages of the case study, can operate with profit also in poorly populated areas and be highly relevant to the livelihood of less mobile people, but also create new opportunities for the consolidation of a sense of local community. For this reason, the creation of new centralities where to supply services in proximity and the incentive to (re)open commercial activities in peripheral areas could represent integrated solutions to guarantee to inhabitants and temporary territory users an opportunity to reach with greater ease services and resources that are now absent and whose access requires long travel out of the proximity sphere. Additionally, in these same areas, economic diversification could involve the implementation of infrastructures that support activities primarily based on connectivity, such as remote working. Evidence suggests that, especially after the Covid-19 pandemic, the number of people working remotely and using coworking spaces—particularly those provided in low-density, inner, and rural areas—has increased (Akhavan et al. 2022). These facilities, often utilizing underused real estate assets (e.g., former schools or libraries no longer in operation), can be significant not only for newcomers but also for the community that revolves around these spaces, providing services to more isolated communities (Bosworth et al. 2023) and creating opportunities for the activation of local economies thanks to the attractiveness of these locations, both as places to visit and as workplaces. However, Mariotti and Di Matteo (2020) warn that the same challenges faced by peri-urban areas, starting with limited digital and physical accessibility, can be significant in influencing the relocation choices of these workers, necessitating careful evaluation of the potential costs and benefits associated with such interventions.

Finally, the fourth category of policies for the x-minute city relevant to immobility enablement concerns local participation and assessment methodologies in evaluating local needs, defining the interventions to be carried out, and monitoring their outcomes. Since strategies for the x-minute city relate to the sphere of proximity, it is crucial that they are developed with consideration of the needs of the different population groups living in the neighborhood. As discussed in Chap. 4, a context-sensitive and aware application of the x-minute city model and immobility enablement requires analyzing the emerging needs of a plurality of stakeholders with different interests and objectives. This attention is important to gather inputs and proposals through listening to local knowledge, to design policies that are at least partly the result of a process of consultation with the population or their representatives, and to calibrate any investigative tools based on local specificities. For example, in Bologna, the so-called "Laboratori di quartiere" have long been active, serving as spaces for listening and dialogue between the administration and citizens on projects and plans affetcing the neighborhood. The information gathered from these exchanges constitutes a fundamental resource for the development of policies that inherently incorporate a local component, which can lead to more effective transformation, ultimately resulting in a stronger relationship between individuals and their neighborhood, also thanks to the attention paid to identifying spaces and interventions aimed at making it more welcoming for inhabitants. The same information can be systematically collected for the construction of assessment methods in both the ex-ante and ex-post phases, through which to evaluate the impact of the interventions and the transformations they have initiated. Similar policies can also be envisaged for peri-urban areas, with the idea of reconstituting community interactions in potentially socially impoverished contexts, thereby creating the conditions for the development or strengthening of relationships among community members.

6.3 Policies for Mobility Enablement

Differently from densely populated urban centers, typically characterized by a strong mobility demand matched by a more or less efficient and diverse supply of public and shared transportation options, in peri-urban areas characterized by lower population and settlement density, it is complex to foresee the widespread diffusion of service facilities that would make them accessible within a short time and distance without the use of a private vehicle. Furthermore, in these contexts, public transportation systems play a marginal role in supporting daily mobility practices and struggle to compete with cars (Berg and Ihlström 2019; Beria 2020). The multiplication and territorial diffusion of activities to ensure accessibility by proximity and induce forms of reversible immobility may not be feasible in every peri-urban context, primarily due to economic considerations linked to economies of scale in service provision. Simultaneously, the limited local availability and diversity of activities may not suffice to meet the manifold and complex accessibility needs arising from different individuals living in the same area.

In these circumstances where proximity cannot always be guaranteed, it is useful to revisit the concept of *accessibility by mobility* (Levine et al. 2019), focusing on the social and economic importance of ensuring essential displacements for people to participate in necessary and desired activities not available nearby. This should be achieved by promoting or incentivizing forms of mobility that are both bottom-up and collaborative, as well as top-down, and that are as efficient and sustainable as possible in social, economic, and environmental terms. We define policies guided by this perspective as mobility enablement policies, which complement immobility enablement in promoting more sustainable and inclusive communities and territories. From a service provision standpoint, these policies particularly focus on offering mobility solutions alternative to the private car that differ from the typical, poorly flexible, and non-personalized configuration of fixed-route public transport and can be applied in both dense and peri-urban contexts, depending on local characteristics, opportunities, and possibilities.

According to the framework proposed by the SMARTA project (2021), alternative mobility solutions to traditional supply can be categorized into three main groups: flexible transport services, ride-sharing, and asset sharing.

The first category, flexible transport services, includes Demand Responsive Transport (DRT) which has long been proven worldwide to be suitable for environments with low mobility demand (Sörensen et al. 2021). DRT schemes have traditionally been structured according to a dial-a-ride service. However, today, the definition of DRT has expanded to include all forms, including digital technology-based, that allow passengers to request transport even with short lead time, both door-to-door and linked to specific routes and stops. Although they are generally subsided with public funds and possibly aimed at target groups with specific mobility needs, they present criteria of flexibility and cost-effectiveness that make them preferable in rural contexts compared to traditional transportation schemes (Mounce et al. 2018). The fact that a DRT system has recently been tested in the case study covered in this research by the Piacenza provincial transit authority provides some insights for assessing the usefulness, effectiveness and transferability, as well as possible short-comings, of these transport solutions when tested in areas with low population and service densities. The Piacenza Apennines DRT experiment developed in the munic-ipalities of Val Nure represents a top-down measure carried out by the provincial transit agency system aimed at guaranteeing to the inhabitants of the most isolated hamlets, generally not served by public transport, a connection with the central core of the municipality. The demand-responsive transit system was introduced starting in 2019 with a minimal number of daily rides. However, this number has increased over the years, shifting from experimental to permanent, suggesting that this measure has likely favored users, as claimed by transportation authority representatives and local politicians interviewed during the fieldwork. However, the system has shown some limitations related in particular to the extent of the area of service that, although defined at the supra-local level, it is still organized into territorial basins that fall within individual municipalities. As a consequence, the service is always routed to the central core of the municipality, even from remote locations that may be closer to other centers. At the same time, residents interviewed on-site showed little famil-iarity with the system, perhaps due to limited advertising of this new transit option in the early stages of experimentation. While not offering exhaustive elements for an overall assessment of the effectiveness of these solutions in rural and mountainous areas, which are still widely discussed in the literature (see Sörensen et al. 2021 for an in-depth account), the experience of the DRT in the Nure valley shows some interesting insights that can be useful to imagine ways of upgrading, promoting and extending the service also in other similar areas. While a DRT system can poten-tially respond to the needs of mobile people depending on how the service is orga-nized regarding flexibility, efficiency of routing, timetabling and booking methods by offering a transportation option closer to the user and their needs (Wang et al. 2015), the lack of territorial wise management and the reference to administrative limits in the prevision of services may make them less adequate to users and efficient, even from the economic point of view. In addition, the information offered to the public, communication and ease of interaction with the service (booking procedures, awareness of how the scheme works, knowledge of costs) become essential elements to encourage users to become familiar with this type of transit scheme, increasing its acceptance and use.

The second type of solution involves ride-sharing, encompassing all forms of trip sharing based on an agreement between the owner of the vehicle and passengers. Among the ride-sharing solutions that hold particular promise in peri-urban contexts are carpooling, volunteer lift-giving, and shared taxi services. Some of these solutions can leverage user collaboration in their organization and management, either informally or through the creation of dedicated digital platforms that enable real-time matching of demand and supply. Various experiences, particularly in the realm of carpooling, demonstrate the potential of these systems, where the development of networks is often shaped by the type of participants and their existing relationships. For example, co-workers may benefit economically by organizing with fewer cars to commute to work, similar to how a family organizes its trips. In any case, carpooling is considered a potential sustainable transport option that can expand mobility opportunities for those without a car, reduce costs for participants, alleviate traffic congestion, and lower emissions by using existing infrastructure more efficiently with minimal need for additional investments (Neoh et al. 2017). Supporting the development of carpooling and ride-sharing systems can thus be adopted as a policy objective by various stakeholders, both public (e.g., urban and peri-urban municipalities) and private (e.g., companies for their employees). However, for such a system to function on a large scale, it is necessary to reach out to potential participants, incentivize them, and encourage sustained carpooling behavior. Several studies have explored the factors that can motivate people to participate in carpooling and ride-sharing schemes. Julagasigorn et al. (2021) introduced a framework that simultaneously considers the relevance of the individual benefits users obtain from using a particular car-pooling service, the importance of relationships within a network that facilitates exchange and interaction and, not least, the acceptance and familiarity with IT technologies that underpin many ride-sharing systems by matching potential users. Understanding these factors is crucial for guiding appropriate implementation strategies by policymakers organizing the systems, especially if designed to include people who are not connected by pre-existing ties or knowledge and is mediated by digital tools and platforms. Thao et al. (2021) particularly emphasize the importance of designing schemes in a participatory way, involving local municipalities and populations. Additionally, functional integration with ordinary transport systems and careful calibration of pricing strategies, if the system involves costs for its use, are essential. Moreover, factors such as system promotion, the introduction of incentives, and continuous monitoring and updating of activities and technological architecture (if present) are key to the success of these systems in various territorial contexts.

Other promising mobility enablement solutions include voluntary lift-giving and shared taxi services. The first category broadly encompasses any form of volunteer-run mobility service in which an individual drives a vehicle, possibly their own, to support others' mobility needs strictly on a non-profit basis. Similarly, other forms of volunteer-led transport involve the use of vehicles not owned by the driver but used to provide the service, as in community transport systems. These transport schemes tend to be small-scale, fully based on voluntary action, targeted at mobility profiles with specific mobility-related impairments, and often subsidized by the public sector (Ravensbergen and Schwanen 2023). They can significantly contribute to the well-being and social participation of less mobile populations. A similar model is seen in shared taxi services, such as the social taxi system in the Piacenza Apennines. The social taxi system, as it is experienced in some of the municipalities in the case study area and, more generally and under different forms and rules, in other peripheral municipalities in Europe, meets the mobility needs of citizens who are not independent in their movements. Such mobility options usually depend on the collaboration of employees and volunteers from local associations who drive a car made available by the municipalities to accompany people who request the service both inside and outside the municipality. In the case study area, the service is present in many villages of the high valleys, although the rules of use, booking procedures, admitted users, rates and hourly availability of the service vary from municipality to municipality. Interviews have shown that residents with reduced autonomy appreciate the service, although a weakness observed relates to the current configuration, which lacks territorial homogeneity and a consistent information and promotion strategy among residents. In this regard, the interviews reveal the opportunity to redesign the services to ensure the same conditions and use across the territory. Nevertheless, shared transport options, including those voluntary and community-based could be extended to larger groups of target users including them in broader territorial strategies aimed at facilitating access to services where this is extremely limited and in cases where more structured transport schemes such as DRT may not adequately respond to the specific and limited demand expressed in dispersed and remote areas.

Another type of mobility solution that has been extensively tested and widely implemented in cities pertains to the third category concerning asset sharing. Car and bike sharing have indeed become highly developed in urban regions as alternative options to private car ownership (Papas et al. 2023). However, this intense development has favored certain areas more than others, with asset sharing systems concentrated in central, densely populated, and socioeconomically privileged areas (Groth et al. 2023). This focus has limited the expansion of such systems into less profitable areas, notably peri-urban environments with medium or low population density, in line with a market-oriented approach that often underpins the organization and management of these systems. Nevertheless, various experiences demonstrate the possibility of creating asset-sharing systems even in peri-urban contexts, for instance, by relying on community-based self-organization. An example of this is the voluntary-based system Der Vaterstettener Autoteiler e.V. (VAT 2024) in the mid-sized German city of Vaterstetten, which operates a non-profit car-sharing system based on membership. Other experiences involve systems organized according to flexible management schedules, ensuring that, during off-peak hours, assets are available for continuous use by businesses or organizations, while being accessible to the public during non-peak hours or on weekends. This type of arrangement is applied in the German cities of Lohr, Karlstadt, and the municipality of Veitshöchheim (Die Energie 2024). Furthermore, asset-sharing systems can be implemented through agreements between sharing providers and private communities, as seen in the case of shared cars for large condominiums or residential communities in the Italian city of Genoa (Elettra 2024). In addition to car sharing, these services can focus on active mobility, allowing users to share traditional or electric bicycles. This type of asset sharing, widely adopted in urban centers, can also be implemented in peri-urban settings by creating territorial networks, as exemplified by the province of Trento in Italy, where a station-based system is deployed across the entire provincial territory with uniform access costs and criteria. Stations are strategically located to serve villages and towns, facilitating intermodal connections with traditional transport services (Bicincittà 2024) by leveraging an efficient network of cycle and pedestrian paths. The system is specifically designed to support last-mile connections between public transport stops and residences, as well as to facilitate local mobility within villages and surrounding areas.

Thus far, the focus has been on the characteristics and implementation possibilities of mobility enablement services as alternatives to traditional public transport, particularly promising for peri-urban contexts. As discussed, some of these systems, including carpooling, can rely on digital technologies for their operation. Indeed, technological advancements in various fields—from IT development to electric motion and vehicle automation—are factors that can significantly influence the social, technical, and economic context in which sustainable and inclusive mobility enablement policies can be designed and implemented. For instance, the development of IT technologies underpins the concept of MaaS (Mobility as a Service), defined as user-centric, multimodal, sustainable, and intelligent mobility management and distribution systems where multiple mobility service providers and end-users interact through digital interfaces (Karmagianni and Goulding 2018). MaaS is not a mobility service per se, but rather creates an environment that facilitates multimodal integration even on a wide territorial scale (e.g. the region, the province or the territory covered by the local transit authority). In fact, not only do MaaS platforms allow users to plan their trips using a variety of transportation systems of a traditional nature (such as ordinary or reservation-based local public transportation), but they also facilitate intermodality with forms of private shared mobility for the last mile (Aapaoja et al. 2017) which can be based on user collaboration. In peri-urban areas, MaaS platforms could be configured as virtual environments in which local mobility demand and supply would be put in contact, facilitating shared travel through forms of car-sharing and pooling and mobility enablement. A transition from individual use of transport to a collective organization would be fostered by encouraging the involvement of the inhabitants of a community in the co-production and management of services (Nunes et al. 2014; Ciasullo et al. 2017). In this perspective, the shift from the concept of MaaS to the concept of Mobility-as-a-Community (MaaC) is proposed in which the functions of MaaS systems are complemented by the possibility of co-producing and co-organizing already existing informal mobility services. This possibility might lay the foundations for an integrated and shared approach to the design and management of (info)mobility platforms and open up scenarios for the reorganization of transport and welfare services at the territorial level (Pucci et al. 2021a).

Further opportunities could also be offered by the progressive electrification of the vehicle fleet, particularly through a more targeted distribution of incentives that many governments provide to individuals for the purchase of low-emission vehicles. As Pucci et al. (2021b) argue, it would be advisable to use public resources to promote the decarbonization of the vehicle fleet in settlement contexts where cars are often essential for last-mile travel, including all those peri-urban areas where, for various reasons, immobility and mobility enablement policies focused on sharing instead of ownership are not feasible. Such a policy could be integrated with measures supporting community-based energy production, such as through the establishment of Positive Energy Districts (JPI Urban Europe 2021), creating net benefits in terms of independence from electric grids still reliant on fossil fuels, thereby enhancing the sustainability of electric vehicles. Resources could also be directed towards the development of sharing systems in denser urban or peri-urban areas, where the promotion and gradual transition from the paradigm of ownership to that of sharing vehicles would be encouraged. Improvements in transport connectivity between urban and peri-urban areas would also facilitate travel to and from dense urban centers, utilizing the potential of privately owned electric vehicles to support last-mile travel to and from public transport hubs.

Finally, the prospect of automation, which can enhance traffic safety, improve the efficiency and flexibility of ordinary transport while reducing costs (Goldbach et al. 2022), could represent an interesting opportunity for mobility enablement, especially for peri-urban contexts, if public acceptance is achieved. According to Silvestri et al. (2022), the potential for automation to make collective transport options more competitive than private cars will materialize if automation meets the needs of those who predominantly use cars for their daily commutes. This will largely depend, according to the authors, on the type of consumption model that will emerge for this technology: if it is oriented towards collective use rather than private ownership, urban and, especially, peri-urban areas could become ideal contexts for the consolidation of flexible, on-demand autonomous transport systems that facilitate last-mile travel to and from public transport nodes and local movements within the territory, thereby overcoming the current gap between public transport and private car use.

In conclusion, the mobility enablement policies and technologies presented here integrates urban and transport planning, urban design, and technological innovations that will be available in the very next future. They should be interpreted as policies to be developed either in an integrated or alternative manner to the immobility enablement policies discussed in the previous section, through an approach that is aware of the characteristics of each context. Indeed, the transfer of policies between different contexts will require adapting the design of interventions to the specific peculiarities, existing actions, and needs of each area of investigation and implementation, which can be assessed through the use of diagnostic tools at both supra-local and more local scales, such as those proposed in the previous chapter of this book.

6.4 Conclusions: From Immobility Analysis to Immobility Enablement

The approach proposed in this book has contributed, through the results obtained herein, to reconsider the role and meaning of forms of relative immobility and daily mobility differentials between individuals in a society built around the assumption of high mobility (Preston and Raje 2007).

The 'negative' view of immobility has probably been fed by the culturally established idea that mobility is in many cases exercised as a privilege and that an ideal good society would not limit physical travel, extending co-presence to every social group and regard any infringement as undesirable (Elliott and Urry 2010). The same authors, foreshadowing possible scenarios of future planetary development, observe how a disruption in current forms of hypermobility induced by shortages of the natural resources needed to set them in motion could create a dramatic shift towards immobile societies (see also Ferreira et al. 2017) that are much more intensely local and smaller in scale. In these new forms of self-reliant and semi-isolated communities, long-distance travel would be uncommon, and lives would again become organized around 'neighborhoods' while planners, politicians, and citizens would collaborate in the redesign of urban and rural centers, neighborhoods, and mobility systems focused upon local low-carbon access and high-level facilities (Elliott and Urry 2010, pp. 143–144).

It is interesting to note that this localist and relatively immobile scenario is proposed by the authors as an unlikely radical consequence of a post-peak oil crisis phase and is somehow defined as less preferable because of the contraction of relationships and networks that mobility helps to generate. However, many elements discussed in this book partly question such a view focusing on those situations in which mobility itself could be coerced and immobility could not. At the same time, the recent interest placed by many cities around the world towards the dimensions of accessibility and proximity, increasingly translated, as discussed, in concrete urban policies, could anticipate a more gradual shift toward a possible reversibly immobile scenario in a less radical way than predicted by Elliott and Urry in 2010. Forms and daily practices of relative immobility or active micro-mobility would not wholly replace mobility and the need for travel but would redefine their geographical scale and intensity. In this perspective, that is particularly promising in terms of increased environmental sustainability of both urban settlements and mobility system, social inclusion will not only be related to the possibility of disposing of a network capital powered by mobile physical interaction, communication and relation at-a-distance, but also (and above all) the ability and possibility to both potentially and actually access these opportunities, regardless of how much this translates into extended, seamless physical travel.

This work offers insights for advancing the debate in planning concerning the main issues related to measuring immobility, identifying spatial conditions that facilitate reversible immobility, and exploring the individual experience of immobility. These issues converge on a fundamental point: the implications of a shift in perspective on immobility, conceived as "*a lens to challenge the grand narrative of hypermobility, flux, and fluidity associated with modernity*" (Schewel 2020, p. 332), for the operational modes and tools of integrated transport and land use planning strategies and measures. The validity of this perspective shift, which positions immobility at the center and considers it, if reversible, as a goal that urban policies should strive for, is supported by the results and insights presented in the various chapters of this book.

Firstly, it has been demonstrated that a lower daily propensity to move is not always to be associated with forms of social and spatial marginality. On the contrary, a wide variety of cases and ways of expressing and perceiving one's practices of immobility beyond linear and defined schemes and relations have been identified. In particular, bridging between a theoretical and operational dimension, this work has underlined how the boundary between chosen and reversible or constrained mobilities and immobilities is not always clear and can be read according to the capacities and possibilities of direct or indirect access to material and immaterial resources. The immobility practices of an individual are the result of the simultaneous action of enabling and coercive factor; such factors are influenced in part by the spatial and functional characteristics of the context and in part by forms of interdependence with mobility and immobility of other individuals, in part by personal preferences, wills, desires and perceptions, creating varied forms of immobility-related *rationalities* and behaviors. Hardly can these rationalities be fully understood in their complexity by referring to established conceptions, values and analytical tools that allocate specific positive attributes to mobility, speed, and acceleration, thereby neglecting the role of stillness, stasis, and slowness in explaining people-specific preferences and behaviors, which in turn can create well-being and a sense of inclusion. However, such complexity does not call into question the role of planning and accessibility improvement policies, since these act on the spatial and functional characteristics of each context and can actively contribute to promoting the reversibility of constrained forms of mobility and immobility in all the multiple circumstances in which they may limit an individual's actual and perceived activity participation and inclusion.

Secondly, the work has deepened the sense of a shift of focus for integrated land use and transport planning from the promotion of mobility to a reversible immobility improvement. At the same time, using a low-density and dispersed mountain territory as a case study, it has illustrated the challenges of adopting such a perspective outside compact urban centers and the need for a comprehensive rethinking of analytical diagnostic tools to assess existing immobility behaviors and accessibility conditions, highlighting the importance of considering a broad spectrum of users and needs in the implementation of measurement tools. The combined use of quantitative and qualitative methods, including direct involvement of local community members and stakeholders through public participation processes, is therefore crucial for calibrating measurement tools and gathering essential insights for designing and implementing mobility and immobility enablement policies tailored to the needs and peculiarities of each location. Research to refine the approach proposed in this book, and exemplified by the framework in Chap. 2, would be in any case fundamental to increase the possibility to identify specific problematic situations, know any spontaneous responses developed by the community to face such problematics, and experiment—or consolidate, if already implemented—specific measures of a local or supra-local nature of reversible immobility and mobility enablement to ensure everyone the right to perform, without compromising their wellbeing, their desired and needed mobility and immobility practices.

References

Aapaoja A, Eckhardt J, Nykänen L, Sochor J (2017) MaaS service combinations for different geographical areas. In: 24th World congress on intelligent transportation systems, Montreal, Canada. https://publications.vtt.fi/julkaisut/muut/2017/OA-MaaSservice-combinations.pdf

Ajuntament de Barcelona (2014) Superillas. Available at https://ajuntament.barcelona.cat/superille s/en/. Accessed 31 July 2024

Akhavan M, Mariotti I, Rossi F (2022) Lo sviluppo degli spazi di coworking nelle aree periferiche e rurali in Italia. TERRITORIO 97:35–42. https://doi.org/10.3280/tr2021-097-supplementoo a12925

Alexander C et al (1987) A new theory of urban design. Oxford University Press, Oxford

Banister D (1995) Transport and urban development. Spon, London

Berg, J., Ihlström J. (2019) The importance of public transport for mobility and everyday activities among rural residents. Soc Sci 8 (2)

Beria P (2020) Quale mobilità durante e dopo il COVID19 nei territori fragili? Blog del Dipartimento di Eccellenza sulle fragilità territoriali. Available at: https://www.eccellenza.dastu.polimi.it/wp-content/uploads/2020/05/Beria-2020-Mobilità-covidterritori-fragili.pdf

Bertolini L (2020) From "streets for traffic" to "streets for people": can street experiments transform urban mobility? Transp Rev 40(6):734–753. https://doi.org/10.1080/01441647.2020.1761907

Bicincittà (2024) Sistema di bike sharing della Provincia di Trento. Available at: https://www.bic incitta.com/frmLeStazioniComune.aspx?ID=187. Accessed 31 July 2024

Büttner B, Seisenberger S, McCormick B, Silva C, Teixeira JF, Papa E, Cao M (2024) Mapping of 15-minute city practices. Overview on strategies, policies and implementation in Europe and beyond. Driving Urban Transition

Bosworth G, Whalley J, Fuzi A, Merrell I, Chapman P, Russell E (2023) Rural co-working: new network spaces and new opportunities for a smart countryside. J Rural Stud 97:550–559. https://doi.org/10.1016/j.jrurstud.2023.01.003

Ciasullo MV, Palumbo R, Troisi O (2017) Reading public service co-production through the lenses of requisite variety. Int J Bus Manag 12(2):1. https://doi.org/10.5539/ijbm.v12n2p1

Comune di Milano (2019) Piano di Governo del territorio. Availbale at https://www.pgt.comune.milano.it. Accessed 31 July 2024

Die Energie (2024) Car sharing In Zusammenarbeit mit den Städten Lohr, Karlstadt und Veitshöchheim. Available at https://die-energie.de/mobilitaet/carsharing/. Accessed 31 July 2024

Dodier R (2013) Modes d'habiter périurbains et intégration sociale et urbaine. EspacesTemps.net, Peer review, 06.05.2013 http://www.espacestemps.net/articles/modes-dhabiter-periurbains-et-integration/

Dorpspunt project Beveren (2024) Presentation of the Dorpspunt project in Beveren (BE). Available at https://delovie.be/dorpspunt-in-beveren/. Accessed 31 July 2024

Elettra (2024) Car sharing di comunità nel Comune di Genova. Available at: https://www.elettraca rsharing.com/comunita/. Accessed 31 July 2024

Elliott A, Urry J (2010) Mobile lives. Routledge, New York

Ferreira A, Bertolini L, Næss P (2017) Immotility as resilience? A key consideration for transport policy and research. Applied Mobilities 2(1):16–31. https://doi.org/10.1080/23800127.2017.1283121

Gehl J (2011) Life between buildings. Island Press, Washington D.C.

Goldbach C, Sickmann J, Pitz T, Zimasa T (2022) Towards autonomous public transportation: attitudes and intentions of the local population. Transp Res Interdiscip Perspect 13:100504. https://doi.org/10.1016/j.trip.2021.100504

Groth S, Klinger T, Otsuka N (2023) Geographies of new mobility services: the emergence of a premium mobility network space. Geoforum 144(103765):5

JPI Urban Europe (2021) Positive energy district. Available at: https://jpi-urbaneurope.eu/ped/. Accessed 31 July 2024

Julagasigorn P, Banomyong R, Grant DB, Varadejsatitwong P (2021) What encourages people to carpool? A conceptual framework of carpooling psychological factors and research propositions. Transp Res Interdiscip Perspect 12:100493. https://doi.org/10.1016/j.trip.2021.100493

Kamargianni M, Goulding R (2018) The mobility as a service maturity index: preparing the cities for the mobility as a service era. In: Proceedings of 7th transport research arena TRA 2018. Zenodo: April 16–19, 2018, Vienna, Austria

Khavarian-Garmsir AR, Sharifi A, Abadi MHH, Moradi Z (2023) From garden city to 15-minute city: a historical perspective and critical assessment. Land 12(2):512. https://doi.org/10.3390/land12020512

Lanza G, Pucci P (2022) Distributing, DesynchroniSing, DigitaliSing: towards a new mobile urbanity in the COVID-19 era. In: Balducci S, Armondi S, Bovo M, Galimberti B (eds) Cities learning from a pandemic: towards preparedness. Routledge

Levine J, Grengs J, Merlin LA (2019) From mobility to accessibility. Transform urban transportation and land use planning. Cornell University Press, Ithaca (NY)

Mariotti I, Di Matteo D (2020) Coworking in emergenza Covid-19: quali effetti per le aree periferiche?EyesReg, vol 10, N 2, Marzo 2020

Masella C, Sgarbossa C (2021) La telemedicina in Italia: cosa è successo durante l'emergenza e cosa fare? Sei azioni per la Sanità del future. Available at: https://www.sanita24.ilsole24ore.com/art/medicina-e-ricerca/2021-05-28/la-telemedicina-italia-cosae-successo-l-emergenza-e-cosa-fare-il-futuro-sei-azioni-la-sanita-futuro-091807.php?uuid=AEAbdYM&refresh_ce=1. Accessed 11 Apr 2022

Mounce R, Wright S, David Emele C, Zeng C, Nelson JD (2018) A tool to aid redesign of flexible transport services to increase efficiency in rural transport service provision. J Intell Transp Syst 22(2). https://doi.org/10.1080/15472450.2017.1410062

Neoh JG, Chipulu M, Marshall A (2017) What encourages people to carpool? An evaluation of factors with meta-analysis. Transportation 44(2):423–447. https://doi.org/10.1007/s11116-015-9661-7

Nunes AA, Galvão T, Cunha JF (2014) Urban public transport service co-creation: leveraging passenger's knowledge to enhance travel experience. Procedia Soc Behav Sci 111(1):577–585. https://doi.org/10.1016/j.sbspro.2014.01.091

Papas T, Basbas S, Campisi T (2023) Urban mobility evolution and the 15-minute city model: from holistic to bottom-up approach. Transp Res Procedia 69:544–551

Preston J, Rajé F (2007) Accessibility, mobility and transport-related social exclusion. J Transp Geogr 15(3):151–160. https://doi.org/10.1016/j.jtrangeo.2006.05.002

Pucci P (2017) Mobility behaviours in peri-urban areas. The Milan urban region case study. Transp Res Procedia 25:4229–4244. https://doi.org/10.1016/j.trpro.2017.05.227

Pucci P, Coppola P, Comai S, Lanza G, Manfredini F (2021a) MAAC_Mobility as a community. Promoting proximity accessibility in fragile territories. Research project

Pucci P, Lanza G, Del Fabbro M (2021b) Superare logiche distributive nei sussidi per la mobilità elettrica e integrare i servizi in un'ottica territoriale. In: Lanzani A, Coppola A, Zanfi F, Pessina G, Del Fabbro M (eds) Ricomporre I divari, politiche e progetti contro le disuguaglianze

Ravensbergen L, Schwanen T (2023) Care-driven informality: the case of community transport. Geogr J 190(2). https://doi.org/10.1111/geoj.12552

Schewel K (2020) Understanding immobility: moving beyond the mobility bias in migration studies. Int Migr Rev 54(2):328–355. https://doi.org/10.1177/0197918319831952

Silvestri F, De Fabiis F, Coppola P (2022) Veicoli a guida autonoma e mobilità post-car. In: Coppola P, Pucci P, Pirlo G (eds) Ottavo Rapporto sulle città Mobilità & Città: verso una post-car city. Il mulino, Bologna

SMARTA project (2021) Final brochure. Available at https://ruralsharedmobility.eu. Accessed 31 July 2024

Soja EW (2011) Regional urbanization and the end of the metropolis era. In: Bridge G, Watson S (eds) New companion to the city. Wiley-Blackwell, Cambridge, Ma, pp 679–689

Sörensen L, Bossert A, Jokinen JP, Schlüter J (2021) How much flexibility does rural public transport need? Implications from a fully flexible DRT system. Transport Policy 100:5–20. https://doi.org/10.1016/j.tranpol.2020.09.005

Stadsbouwmeester Gent (2020) Stadsgebouw 2.0 final report. Available at: https://stad.gent/sites/default/files/media/documents/SBMG-ESSAY%20Stadsgebouw%202.0_A5_Binnen_FINAL_LR.pdf. Accessed 31 July 2024

Thao VT, Imhof S, Von Arx W (2021) Integration of ridesharing with public transport in rural Switzerland: Practice and outcomes. Transp Res Interdiscip Perspect 10:100340. https://doi.org/10.1016/j.trip.2021.100340

VAT (2024) Car sharing in Vaterstetten. Available at https://www.carsharing-vaterstetten.de. Last access 2024/07/31

Ville de Paris (2022) Le label fabriqué a Paris cours oasis. Available at https://www.paris.fr/pages/le-label-fabrique-a-paris-5152. Accessed 31 July 2024

Ville de Paris (2023) Les cours oasis. Available at: https://www.paris.fr/pages/les-cours-oasis-7389. Accessed 31 July 2024

Wang C, Quddus M, Enoch M, Ryley T, Davison L (2015) Exploring the propensity to travel by demand responsive transport in the rural area of Lincolnshire in England. Case Stud Transp Policy 3(2):129–136. https://doi.org/10.1016/j.cstp.2014.12.006